I0055072

Unification of the Strong Interactions and Gravitation

Quark Confinement Linked to Modified Short-Distance Gravity

Physics is Logic VIII

STEPHEN BLAHA

BLAHA RESEARCH

ISBN: 978-0-9970761-4-1

Cover Credits

Rev. 00/00/01 June 7, 2016

To My Wife Margaret

Some Other Books by Stephen Blaha

All the Megaverse! Starships Exploring the Endless Universes of the Cosmos using the Baryonic Force (Blaha Research, Auburn, NH, 2014)

All the Universe! Faster Than Light Tachyon Quark Starships & Particle Accelerators with the LHC as a Prototype Starship Drive Scientific Edition (Pingree-Hill Publishing, Auburn, NH, 2011).

From Asynchronous Logic to The Standard Model to Superflight to the Stars; Volume 2: Superluminal CP and CPT, U(4) Complex General Relativity and The Standard Model, Complex Vierbein General Relativity, Kinetic Theory, Thermodynamics (Blaha Research, Auburn, NH, 2012)

The Algebra of Thought & Reality: The Mathematical Basis for Plato's Theory of Ideas, and Reality Extended to Include A Priori Observers and Space-Time; Second Edition (Pingree-Hill Publishing, Auburn, NH, 2009)

SuperCivilizations: Civilizations as Superorganisms (McMann-Fisher Publishing, Auburn, NH, 2010)

Universes and Megaverses: From a New Standard Model to a Physical Megaverse; The Big Bang; Our Sister Universe's Wormhole; Origin of the Cosmological Constant, Spatial Asymmetry of the Universe, and its Web of Galaxies; A Baryonic Field between Universes and Particles; Flatverse Extended Wheeler-DeWitt Equation (Blaha Research, Auburn, NH, 2014)

PHYSICS IS LOGIC PAINTED ON THE VOID: Origin of Bare Masses and The Standard Model in Logic, U(4) Origin of the Generations, Normal and Dark Baryonic Forces, Dark Matter, Dark Energy, The Big Bang, Complex General Relativity, A Megaverse of Universe Particles (Blaha Research, Auburn, NH, 2015).

PHYSICS IS LOGIC Part II: The Theory of Everything, The Megaverse Theory of Everything, U(4)⊗U(4) Grand Unified Theory (GUT), Inertial Mass = Gravitational Mass, Unified Extended Standard Model and a New Complex General Relativity with Higgs Particles, Generation Group Higgs Particles (Blaha Research, Auburn, NH, 2015).

The Origin of Higgs ("God") Particles and the Higgs Mechanism: Physics is Logic III, Beyond Higgs – A Revamped Theory With a Local Arrow of Time, The Theory of Everything Enhanced, Why Inertial Frames are Special, Universes of the Mind (Blaha Research, Auburn, NH, 2015).

The Origin of the Eight Coupling Constants of The Theory of Everything: U(8) Grand Unified Theory of Everything (GUTE), S^8 Coupling Constant Symmetry, Space-Time Dependent Coupling Constants, Big Bang Vacuum Coupling Constants, Physics is Logic IV (Blaha Research, Auburn, NH, 2015).

New Types of Dark Matter, Big Bang Equipartition, and A New U(4) Symmetry in the Theory of Everything: Equipartition Principle for Fermions, Matter is 83.33% Dark, Penetrating the Veil of the Big Bang, Explicit QFT Quark Confinement and Charmonium, Physics is Logic V (Blaha Research, Auburn, NH, 2015).

The Periodic Table of the 192 Quarks and Leptons in The Theory of Everything: The U(4) Layer Group, Physics is Logic VI (Blaha Research, Auburn, NH, 2015).

New Boson Quantum Field Theory, Dark Matter Dynamics, Dark Matter Fermion Layer Mixing, Genesis of Higgs Particles, New Layer Higgs Masses, Higgs Coupling Constants, Non-Abelian Higgs Gauge Fields, Physics is Logic VII (Blaha Research, Auburn, NH, 2015)

Available on Amazon.com, Amazon.co.uk, bn.com, and other international web sites as well as at better bookstores (through Ingram Distributors).

Preface

This technical book introduces an extension to our Theory of Everything in which we unify the Strong Interaction and Gravitation sectors. The basic rationale for this unification is the simple fact that the Strong Interactions and General Relativity are the only interactions without broken symmetry, and the closeness of the strength of the Strong Interaction and Gravitation when properly viewed. This higher derivative theory yields quark (color) confinement explicitly with a linear potential. It is consistent with the Charmonium spectrum potential of the Cornell group. The gravitation sector yields the usual Newtonian potential plus two very massive gravitons (of mass 0.119/sqrt(G) and 0.152/sqrt(G) where G is Newton's constant) that modify the extreme short distance (Planck length distance) behavior of gravity. Gravity is initially complex in this unification. The introduction of scalar (Higgs) bosons enables us to define U(4) Reality group transformations to reduce complex General Relativity to the real General Relativity that we observe in Nature.

The unified theory theory directly generates a complex 16-dimensional space (using a vierbein formalism) that we have previously called the Megaverse. Thus our previous studies of the Megaverse, the home of universes, are further supported.

CONTENTS

1. Introduction

The nature of the theory of the Strong Interactions has been a subject of speculation for many years. It is generally believed that it is a Yang-Mills SU(3) gauge theory. The theory is also expected to have color confinement and asymptotic freedom. While there is evidence for these features, they have not been rigorously proved – not even in the physicists' sense of 'rigor.'

General Relativity also has significant issues when attempts are made to quantize it. Here again there are expectations about Quantum General Relativity but there are many unanswered questions.

We juxtapose the theories of the Strong Interaction and Quantum General Relativity because they share a unique feature: both theories do *not* have symmetry breaking. In contrast the theory of ElectroWeak interactions is both well defined and has explicit symmetry breaking features. We also note that the gravitational coupling constant and the Strong Interaction coupling constant, when suitably expressed using pseudoquantum fields, have comparable values.[1] These similarities have led us to propose that the Strong Interactions may be intimately related to the geometry of the universe.[2]

Taking up the issues raised in reference 2 below we will now develop a *vierbein* (tetrad) unified theory of the Strong Interactions and Gravitation that can be viewed as a modified sector within our Theory of Everything. (See Blaha (2015a) and other books by the author.) A preliminary discussion of this unification was presented in Blaha (2016a).

We will begin by developing an overall framework for a unified theory using a complex-valued[3] generalization of the vierbein formalism to include not only a space-time vierbein part but also an SU(3) color gauge part. Then we will establish a theory that effectively factors the vierbeins into a product of a space-time vierbein and an SU(3) part.

The factored parts will be separately considered. The SU(3) part will lead to a confining Strong Interaction theory that is consistent with the generally accepted charmonium potential. The confinement is explicit through the appearance of a linear r potential. The Gravitation part leads ultimately to dynamic equations with a Modified Newtonian Gravity at ultra-short distances.

One important result of this theory is the appearance of a complex 16-dimensional space that we previously called the Megaverse (Blaha (2015a) and more recent books).

---

[1] See Blaha (2015d).

[2] S. Blaha, "Quantum Gravity and Quark Confinement" (March, 1976) Gravity Research Foundation Essay Contest – Honorable Mention. This paper is reproduced in Appendix A.

[3] In complex space-time.

2. The Strong Interaction and Charmonium

As a preliminary to the development of our unified theory of Gravitation and the Strong Interaction we will consider the Strong Interaction potential used in studies of Charmonium bound states.[4] This potential will be seen to be that of our Strong Interaction theory developed some years ago with one simple modification. Subsequently we will develop the structure of the unified theory in some detail.

In 1974 a bound state of a charmed and an anti-charmed quark was discovered by two experimental groups. Since charmed quarks are quite massive, theoretical attempts were made to understand the charmed quark bound states within the framework of non-relativistic quantum mechanics. The "Cornell group" developed a fairly satisfactory[5] charmed quark bound state spectrum in 1974-5 using a combination of a linear and a 1/r potential as the strong interaction. In a recent fit[6] they gave the potential energy:

$$V(r) = -\kappa/r + r/a^2 \qquad (2.1)$$

where $\kappa = 0.61$, $a = 2.38\ \text{GeV}^{-1}$ and the charmed quark mass was 1.84 GeV.

2.1 The Origin of the Linear Potential

The linear potential appears to have originated in a suggestion of Feynman in the Spring of 1974. This author proposed[7] a non-Abelian gauge quantum field theory, which yielded a linear potential. These papers, which had 4th order dynamic equations for the gauge fields, showed how to avoid the problems previously associated with higher derivative theories by using principal-value gauge field propagators that were similar in concept to the action-at-a-distance propagators used by Feynman and Wheeler in the late 1940's to formulate action-at-a-distance Electrodynamics.

Thus we created a non-Abelian quantum field theory of the strong interaction yielding a linear potential. In parallel with this development, Professor Kenneth Wilson (later a Nobelist) was developing lattice gauge theory. Because lattice lines focus the field of gauge bosons, lattice gauge theory also exhibited a linear potential between quarks. Thus it offered an

---

[4] This chapter first appeared in Blaha (2016a).
[5] As did a Harvard group.
[6] E. J. Eichten, K. Lane, and C. Quigg, arXiv:hep-ph/ 0206018 (2002). See this paper for references to earlier work by the "Cornell group" and the "Harvard group" as well as papers by other researchers.
[7] S. Blaha, Phys. Rev. D**10**, 4268 (July, 1974) and Phys. Rev. D**11**, 2921 (December, 1974). They appear in Appendices A and B of this book for the reader's convenience. These papers appeared before the charmonium calculations of the Cornell and Harvard groups in 1975.

alternative to our gauge theory. However, the linearity of the lattice potential was "built-in" by the lattice theory formulation and thus was an artifact of the lattice formulation. This approach and other proposed approaches all share the problem that the linear potential that they produce cannot be proven to truly be a consequence. Rather the linear potential is the "likely result."

On the other hand our higher dimensional theory produces the linear potential if the standard rules of quantum field theory are followed with the proviso that gauge field propagators are principal-value propagators.

This author had many discussions with Professor Wilson in late 1974 in the author's office and while walking to lunch at the Cornell Faculty Club. Professor Wilson proposed possible flaws in the author's theory on an almost daily basis. The author was able to show these possible flaws were not flaws but physically acceptable. The final discussion with Wilson ended with Wilson stating words to the effect, "Your theory may be a correct phenomenological approximation to my theory of the strong interaction and quark confinement. But my theory is the correct one. Your theory is only a phenomenology." In the forty plus years since this concluding discussion no one has proved that the conventional strong interaction theory truly has a linear potential and quark confinement although some approximations suggest it does.

In the absence of a demonstration of a linear potential in the standard strong interaction model we suggest our theory is a viable alternative. Since the linear potential appears to fairly successfully describe much of the charmonium spectrum we feel our theory with its explicit derivation of a linear potential is worthy of interest – especially because it is in agreement with experiment as far as we know. *An experimentally completely correct phenomenology is a theory.*

2.2 Charmonium Potential in the Light of our Theory

The author's non-abelian strong interaction theory presented in Appendices A and B needs one modification to yield the charmonium potential eq. 2.1. In Appendix A eqs. 6 and 18 should have the interaction term expanded:

$$g\mathcal{A}_1 \rightarrow g(\mathcal{A}_1 + \mathcal{A}_2) \tag{2.2}$$

and similarly in eq. 20 in Appendix B.

Eqs. 38 – 41 of Appendix A directly show that the additional interaction term leads to a gluon propagator[8] $\langle A^1 + A^2, A^1 + A^2\rangle = 2\langle A^1, A^2\rangle + \langle A^1, A^1\rangle$, and introduces a 1/r term in the potential part of the propagator. This leads to a potential of the form of eq. 2.1 with $g = \sqrt{(\kappa/2)} = 0.552$ and $\lambda = 0.761$ GeV. Thus

$$V(r) = -2g^2/r + g^2\lambda^2 r \tag{2.3}$$

The charmonium potential emerges directly from our theory. We note that $g^2/4\pi = 0.024$ is only a factor of 3.3 more than the fine structure constant – approximately 1/137 = 0.0073. *Therefore*

[8] Eqs. 40-41 in Appendix A.

perturbative corrections to our inter-quark potential may not be significant and our theory may be the correct theory of the strong interaction. The strong interaction potential in this charmonium fit suggests that the unperturbed potential of the theory presented in Appendices A and B may be a good approximation to the exact potential determined in perturbation theory.

The smallness of the Strong Interaction terms in the Cornell group potential (eq. 2.1) looks puzzling at first glance. Why is it not large ("Strong")? We believe the strength of the Strong Interaction does not originate in the value of the coupling constant but rather in the linear potential term which provides confinement. The Cornell group potential's 'small' coupling constant then becomes understandable within the context of Strong Interaction phenomenology.

2.3 Asymptotic Freedom in Our Model of the Strong Interaction

The combined gluon propagator in momentum space in our theory has the form

$$-1/k^2 + \lambda^2/k^4 \tag{2.4}$$

In the high energy (short distance) limit k → ∞ it approaches k^{-2} yielding the same short distance behavior as the usual strong interaction model. In the low energy (long distance) limit k → 0 it has a confining k^{-4} behavior. Thus we explicitly obtain quark confinement and asymptotic freedom in our model.

3. Why Unite the Strong Interaction and Gravity?

3.1 The Strong Interaction and Gravitation are the Only Unbroken Symmetries

One suggestive qualitative reason for combining the Strong Interaction and Gravitation in a unified sector within the Theory of Everything is their uniqueness as unbroken symmetries. The ElectroWeak interaction has a broken symmetry. The U(4) Generation group symmetry and the U(4) Layer group symmetry are also broken. The Strong Interaction and Gravitation then stand out as unbroken symmetries. Further, the symmetry of space-time encourages the view that these unbroken symmetries are associated with space-time geometry – not just in four dimensions but in a wider space that includes 4-dimensional space-time but also contains color confined 'bubbles' of a 32-dimensional real space, which is equivalent to an 16-dimensional complex-valued space. This 16-dimensional space would be confined to hadrons and the quark-gluon plasmas being created in experiments at CERN and Brookhaven.

3.2 Similarity of the PseudoQuantum Vacuum Expectation Values for Gravity and the Strong Interaction

The Strong interaction and gravity have approximately equal vacuum expectation values generating their coupling constants as shown in eq. 1.43 in Blaha (2015d). For the reader's convenience we reproduce a slightly edited parts of Blaha (2015d) and (2016c):

Beginning of Extract

In Blaha (2015c) we introduced a new formalism for the generation of non-zero vacuum expectation values using a form of second quantization, called pseudoQuantization, developed by the author[9] in 1978 that combines both quantum and classical fields within the same framework. In this extended quantum field theory vacuum expectation values are due to coherent vacuum ground states. This approach has the advantage of 1) resolving the issue of negative energy boson states that has existed for many years;[10] and 2) showing why inertial reference frames are of special significance – a long standing question in Theoretical Physics.[11]

---

[9] This section is based on our 1978 paper that appeared in the peer-reviewed journal *Physical Review D*. The paper is reproduced in Appendix B for the reader's convenience with the kind permission of The American Physical Society. The paper also does present a new formulation of Quantum Mechanics that incorporates both quantum and classical mechanics within one framework. Recently, experimenters have been investigating the possibility of macroscopic quantum phenomena. The new formulation is ideally suited for tracing the change from a quantum to a classical regime. It also is applicable to "large n atoms" where the outermost electrons approach classical behavior with an almost continuous energy spectrum.

[10] P. A. M. Dirac once claimed to have a solution for this problem but he did not reveal it. See Blaha (2015c).

[11] Their significance follows from the existence of a common rest frame for all Higgs particles vacuum states. All other inertial reference frames result from Lorentz transformations. See Blaha (2015c),

1.4.1 PseudoQuantization of Higgs Particles

We will now consider the pseudoQuantization of a scalar particle field that will become a particle with a non-zero vacuum expectation value. (Section III of our paper in Appendix B contains additional detail.) We begin by defining two fields that correspond to a single scalar particle: $\varphi_1(x)$ and $\varphi_2(x)$.[12] These fields will be assumed to have the equal time commutators

$$[\varphi_i(x), \pi_j(y)] = i(1 - \delta_{ij})\delta^3(\mathbf{x} - \mathbf{y}) \quad (1.1)$$
$$[\varphi_i(x), \varphi_j(y)] = 0$$
$$[\pi_i(x), \pi_j(y)] = 0$$

where δ_{ij} is the Kronecker δ and where $\pi_i(x)$ is the canonically conjugate momentum to $\varphi_i(x)$. The fields $\varphi_1(x)$ and $\pi_1(y)$ will be observable classical fields as shown by eqs. 69 and 70 in Appendix B. The fields $\varphi_2(x)$ and $\pi_2(y)$ will not be observables so that $\varphi_1(x)$ and $\pi_1(y)$ can both be sharp on the set of physical states.

We now specify the lagrangian density for a scalar Klein-Gordon particle:

$$\mathcal{L} = \partial\varphi_1/\partial x_\mu \partial\varphi_2/\partial x^\mu - m^2\, \varphi_1\varphi_2 \quad (1.2a)$$

and hamiltonian density

$$\mathcal{H} = \pi_1\, \pi_2 + \partial\varphi_1/\partial x_i \partial\varphi_2/\partial x^i + m^2\, \varphi_1\varphi_2 \quad (1.2b)$$

where i labels spatial coordinates, m is the particle mass,[13] and $\pi_1 = \partial\varphi_2/\partial t$ and $\pi_2 = \partial\varphi_1/\partial t$. The fields can be fourier expanded in terms of creation and annihilation operators:

$$\varphi_i(\mathbf{x}, t) = \int d^3k\, [a_i(k)f_k(x) + a_i^\dagger(k)f_k^*(x)] \quad (1.3)$$

for i = 1, 2 where

$$f_k(x) = e^{-ik\bullet x}/(2\omega_k(2\pi)^3)^{1/2}$$

with ω_k being the energy.

The creation and annihilation operators satisfy the commutation relations:

$$[a_i(k), a_j^\dagger(k')] = (1 - \delta_{ij})\delta^3(\mathbf{k} - \mathbf{k'}) \quad (1.4)$$
$$[a_i(k), a_j(k')] = 0$$
$$[a_i^\dagger(k), a_j^\dagger(k')] = 0$$

for i, j = 1, 2.

In this formulation the defining properties of a coherent physical state are:

$$\varphi_1(x)|\Phi, \Pi> = \Phi(x)|\Phi, \Pi> \quad (1.5)$$
$$\pi_1(x)|\Phi, \Pi> = \Pi(x)|\Phi, \Pi>$$

where $\Phi(x)$ and $\Pi(x)$ are sharp on the states and thus classical fields expressible as

$$\Phi(\mathbf{x}, t) = \int d^3k\, [\alpha(k)f_k(x) + \alpha^*(k)f_k^*(x)] \quad (1.6)$$

and correspondingly for $\Pi(x)$.

1.4.2 Vacuum States for Scalar Particles with Non-Zero Vacuum Expectation Values

When we implement the mass mechanism Φ becomes a constant. We can define a set of states

---

[12] The subscripts on the fields are not gauge symmetry indices but simply identifiers distinguishing the fields from one another.

[13] Note the mass term here has the "correct" sign unlike the mass term in the usual Higgs potential.

$$a_1(k)|\alpha\rangle = \alpha(k)|\alpha\rangle$$
$$a_1^\dagger(k)|\alpha\rangle = \alpha^*(k)|\alpha\rangle$$

and correspondingly a set of coherent states

$$|\alpha\rangle = C\exp\{\int d^3k\,[\alpha(k)a_2^\dagger(k) + \alpha^*(k)a_2(k)]\}|0\rangle \tag{1.7}$$

where C is a normalization constant, and where the vacuum state $|0\rangle$ satisfies

$$a_1(k)|0\rangle = a_1^\dagger(k)|0\rangle = 0 \tag{1.8a}$$
$$a_2(k)|0\rangle \neq 0 \qquad\qquad a_2^\dagger(k)|0\rangle \neq 0 \tag{1.8b}$$

The dual vacuum state satisfies

$$\langle 0|a_2(k) = \langle 0|a_2^\dagger(k) = 0 \tag{1.9a}$$
$$\langle 0|a_1(k) \neq 0 \qquad\qquad \langle 0|a_1^\dagger(k) \neq 0 \tag{1.9b}$$

Additional details on these coherent states, which differ from conventional coherent states such as those of Kibble[14] and others, can be found in Appendix C.

With this coherent state formalism, which gives purely classical fields and yet also has quantum fields through the use of φ_2 and its creation and annihilation operators, we now have the machinery to define a mass mechanism without the introduction of a potential whose origin can only be described as dubious.

For we can define a coherent state for some k as

$$|\Phi, \Pi\rangle = C\exp\{[(2\pi)^3\omega_k/2]^{1/2}\Phi[a_2^\dagger(k) + a_2(k)]\}|0\rangle \tag{1.10}$$

where C is a normalization constant, that yields a constant, non-zero vacuum expectation value:

$$\varphi_1(x)|\Phi, \Pi\rangle = \Phi|\Phi, \Pi\rangle \tag{1.11}$$

where Φ is a constant. Evaluating a fermion interaction term we find a mass term emerges[15]

$$\bar{\psi}(\varphi_1 + \varphi_2)\psi \rightarrow \bar{\psi}(\Phi + \varphi_2)\psi \tag{1.12}$$

It can generate a mass for an interaction with a gauge field of the form

$$A^\mu(\varphi_1 + \varphi_2)^2 A_\mu \rightarrow A^\mu(\Phi + \varphi_2)^2 A_\mu \tag{1.13}$$

It also yields a quantum field theoretic interaction that would result in the production of ElectroWeak particles from these scalar fields. (The production of Higgs particles that decay into ElectroWeak gauge particles has recently been found at CERN.)

The present formalism provides a clean way to separate the vacuum expectation value of a scalar particle from its quantum field part in contrast to the Higgs Mechanism where one has to separate a Higgs field into parts manually.

It appears that our formulation of the mass generation mechanism sheds significant light on the reason for the special prominence of inertial frames. Consider massive scalars.[16] Eqs. 1.2 describe a massive scalar particle. If

---

[14] T. W. B. Kibble, Jour. Math. Phys. **2**, 212 (1961).

[15] When matrix elements with a "vacuum state" such as eq. 1.10 are taken.

[16] Experiments at CERN have apparently discovered a Higgs particle with a 125 GeV/c mass.

the scalar is massive, then the rest frame particle "vacuum" state, eq. 1.10, which yields a non-zero expectation value, is

$$|\Phi, \Pi> = C\exp\{[(2\pi)^3 m/2]^{1/2}\Phi[a_2^\dagger(\mathbf{0},m) + a_2(\mathbf{0},m)]\}|0> \tag{1.10'}$$

We thus find that inertial reference frames are singled out as "special" in the sense that they are the only accessible reference frames that can be generated by a Lorentz boost/transformation from the Higgs particle rest frame. *The particles vacuum state single out the class of inertial reference frames.*

...

7.1 Coupling Constants Vacuum Expectation Value Generation

The appearance of just eight fundamental coupling constants in The Theory of Everything makes them ideal candidates for replacement by eight scalar Higgs particle, vacuum expectation values. *Using vacuum expectation values leads to the remarkable conclusion that all known coupling constants, properly rewritten using our pseudoQuantum vacuum expectation value formalism, all have a value of the order of unity – even the gravitational constant* $g_{CG} = \kappa^{-1} = (4\pi G)^{-1/2}$.

7.1.1 Yang-Mills Coupling Constants

We will first consider the case of a generic Yang-Mills field $A^{b\mu}$ of some symmetry group, a generic fermion field ψ, and a Higgs particle with fields φ_1 and φ_2 (as defined earlier). We will replace its generic coupling constant g with Higgs fields' vacuum expectation values.[17] The initial dynamic equations are

$$\partial/\partial x_\mu\, F^a{}_{\mu\nu} + gf^{abc}A^{b\mu}\, F^c{}_{\mu\nu} = j^a{}_\nu \tag{7.1}$$

and

$$[i\gamma^\mu(\partial/\partial x^\mu - igA_\mu) - m]\psi(x) = 0 \tag{7.2}$$

where

$$F^a{}_{\mu\nu} = \partial/\partial x^\nu A^a{}_\mu - \partial/\partial x^\mu A^a{}_\nu + gf^{abc}A^b{}_\mu A^c{}_\nu \tag{7.3}$$

and where a, b, c are structure constant indices, g is the coupling constant, and $j^a{}_\nu$ is the corresponding current.

A gauge transformation has the form

$$gA'_\mu(x) = -i(\partial_\mu\Omega(x))\Omega^{-1}(x) + g\Omega(x)A_\mu(x)\Omega^{-1}(x) \tag{7.4}$$

7.1.2 C-Number Field Coupling Constant

We can replace g with fields in two ways. One way is:

$$\partial/\partial x_\mu\, F^a{}_{\mu\nu} + m'^{-1}\varphi_1(x)f^{abc}A^{b\mu}\, F^c{}_{\mu\nu} = j^a{}_\nu \tag{7.5}$$

and

$$[i\gamma^\mu(\partial/\partial x^\mu - im'^{-1}\varphi_1(x)A_\mu) - m]\psi(x) = 0 \tag{7.6}$$

where

$$F^a{}_{\mu\nu} = \partial/\partial x^\nu A^a{}_\mu - \partial/\partial x^\mu A^a{}_\nu + m'^{-1}\varphi_1(x)f^{abc}A^b{}_\mu A^c{}_\nu \tag{7.7}$$

with the corresponding gauge transformation rule:

---

[17] This approach is conceptually similar to that of Dicke et al for the gravitational constant G. See R. H. Dicke, Phys. Rev. **125**, 2163 (1962) and references therein. See Weinberg (1972) p. 155ff, and Misner et al (1973) p. 1070 for lucid discussions.

$$\varphi_1(x)A'_\mu(x) = -im'(\partial_\mu\Omega(x))\Omega^{-1}(x) + \varphi_1(x)\Omega(x)A_\mu(x)\Omega^{-1}(x) \quad (7.8)$$

Using the vacuum state defined by

$$|\Phi, \Pi> = C\exp\{[(2\pi)^3 m/2]^{1/2} m'g[a_2^\dagger(\mathbf{0},m) + a_2(\mathbf{0},m)]\}|0> \quad (7.9)$$

we see the equations become eqs. 7.1-7.4 since $\Phi = m'g$. Note m' may equal m, or may be the iota Landauer mass[18] or some other value.

Thus we have developed the first Higgs-like mechanism for purely c-number coupling constants.

7.1.3 Q-Number Field Coupling Constant

The other way to reduce coupling constants to vacuum expectation values, which we believe is preferable, is

$$\partial/\partial x_\mu F^a_{\mu\nu} + m'^{-1}(\varphi_1 + \varphi_2)f^{abc}A^{b\mu}F^c_{\mu\nu} = j^a_\nu \quad (7.10)$$

and

$$[i\gamma^\mu(\partial/\partial x^\mu - i\, m'^{-1}(\varphi_1 + \varphi_2)A_\mu) - m]\psi(x) = 0 \quad (7.11)$$

where

$$F^a_{\mu\nu} = \partial/\partial x^\nu A^a_\mu - \partial/\partial x^\mu A^a_\nu + m'^{-1}(\varphi_1 + \varphi_2)f^{abc}A^b_\mu A^c_\nu \quad (7.12)$$

with the corresponding *q-number* gauge transformation rule:

$$(\varphi_1(x) + \varphi_2(x))A'_\mu(x) = -im'(\partial_\mu\Omega(x))\Omega^{-1}(x) + (\varphi_1(x) + \varphi_2(x))\Omega(x)A_\mu(x)\Omega^{-1}(x) \quad (7.13)$$

Using the vacuum state eq. 7.9 we find, for *real-valued* coordinates, eqs. 7.10 – 7.13 become

$$\partial/\partial x_\mu F^a_{\mu\nu} + (g + m'^{-1}\varphi_2)f^{abc}A^{b\mu}F^c_{\mu\nu} = j^a_\nu \quad (7.14)$$

and

$$[i\gamma^\mu(\partial/\partial x^\mu - i(g + m'^{-1}\varphi_2)A_\mu) - m]\psi(x) = 0 \quad (7.15)$$

where

$$F^a_{\mu\nu} = \partial/\partial x^\nu A^a_\mu - \partial/\partial x^\mu A^a_\nu + (g + m'^{-1}\varphi_2)f^{abc}A^b_\mu A^c_\nu \quad (7.16)$$

with the corresponding *q-number* gauge transformation rule:

$$(gm' + \varphi_2(x))A'_\mu(x) = -i\, m'(\partial_\mu\Omega(x))\Omega^{-1}(x) + (gm' + \varphi_2(x))\Omega(x)A_\mu(x)\Omega^{-1}(x) \quad (7.17)$$

7.1.4 C-Number Coupling Constants or Q-Number Coupling Constants?

The above two possible methods for reducing coupling constants to Higgsian vacuum expectation values have different experimental implications. In the case of ElectroWeak gauge fields a c-number coupling constant does not introduce a new interaction with a Higgs particle. In the case of ElectroWeak gauge fields a q-number coupling constant, it does introduce a new interaction with a new Higgs particle. The Higgs particle found at CERN LHC may have been produced from an ElectroWeak gauge field with a q-number coupling constant term.

7.1.5 Strong Interaction Case for Complex-Valued Complexon Coordinates

In the case of the *strong SU(3)* gauge fields the q-number approach would lead to a new interaction of gluons and a Higgs particle corresponding to the field φ_2.

---

[18] The iota mass is a universal mass equal to the Landauer energy of a logical value. See Blaha (2015a).

The new dynamic equations for the complexon Yang-Mils field upon replacement of the coupling constant g by a Higgs field using our pseudoQuantization formalism are:

$$D_\mu F^a_{\mu\nu} + (g + m'^{-1}\varphi_2) f^{abc} A^{b\mu} F^c_{\mu\nu} = j^a_\nu \tag{7.18}$$

and

$$[i\gamma^\mu(D_\mu - i(g + m'^{-1}\varphi_2)A_\mu) - m]\psi(x) = 0 \tag{7.19}$$

where

$$F^a_{\mu\nu} = D_\nu A^a_\mu - D_\mu A^a_\nu + (g + m'^{-1}\varphi_2) f^{abc} A^b_\mu A^c_\nu \tag{7.20}$$

where all coordinates are *complex-valued* $x = x_r + ix_i$ with derivatives D_μ given by

$$D_0 = \partial/\partial x^0$$
$$D_k = \partial/\partial x^k + i\,\partial/\partial x_i^k$$

with $x^k = x_r^k$ for k = 1, 2, 3.

The corresponding *q-number* gauge transformation rule:

$$(gm' + \varphi_2(x))A'_\mu(x) = -i\,m'(\partial_\mu\Omega(x))\Omega^{-1}(x) + (gm' + \varphi_2(x))\Omega(x)A_\mu(x)\Omega^{-1}(x) \tag{7.21}$$

Q-number gauge transformations appear in a number of situations. For example, Quantum Electrodynamics has q-number gauge transformations.

7.2 Gravitational Coupling Constant Vacuum Expectation Value Generation

The gravitational coupling constant $g_{CG} = \kappa^{-1} = (4\pi G)^{-1/2}$ appears in the gravitational lagrangian density. An example is the case of an interaction with a Dirac particle:

$$\mathcal{L} = g_{CG}^2 \sqrt{g}\, R/2 + a\bar\psi(i\gamma^\mu\partial/\partial x^\mu - m)\psi \tag{7.22}$$

where a is a coupling constant. and g_{CG} has the dimension of mass. Thus we can introduce a coherent vacuum state

$$|\Phi_G, \Pi_G\rangle = C\exp\{[(2\pi)^3 m/2]^{1/2} g_{CG}[a_2^\dagger(\mathbf{0},m) + a_2(\mathbf{0},m)]\}|0\rangle \tag{7.23}$$

similar to eq. 7.9 that enables us to re-express eq. 7.22 as

$$\mathcal{L} = \varphi_1^2 \sqrt{g}\, R/2 + a\bar\psi(i\gamma^\mu\partial/\partial x^\mu - m)\psi \tag{7.24}$$

or

$$\mathcal{L} = (\varphi_1 + \varphi_2)^2 \sqrt{g}\, R/2 + c\bar\psi(i\gamma^\mu\partial/\partial x^\mu - m)\psi \tag{7.25}$$

using the formalism of section 7.4 with the vacuum state eq. 7.23 throughout.[19] Thus we can directly embody the gravitational constant within our formalism.

If we add the pseudoQuantum fields' lagrangian to the lagrangian of eq. 7.24 we obtain:

$$\mathcal{L} = \varphi_1^2 \sqrt{g}\, R/2 + a\bar\psi(i\gamma^\mu\partial/\partial x^\mu - m)\psi + \partial\varphi_1/\partial x_\mu \partial\varphi_2/\partial x^\mu - m^2\varphi_1\varphi_2 \tag{7.26}$$

The dynamic equations for φ_1 and φ_2 are

$$\Box\varphi_2 + m^2\varphi_2 - \varphi_1\sqrt{g}\, R = 0 \tag{7.27}$$

and

$$\Box\varphi_1 - m^2\varphi_1 = 0$$

---

19 Other variants of these equations are possible such as using the term $g_{CG}\varphi_1\sqrt{g}R/2$ instead of $\varphi_1^2\sqrt{g}R/2$.

In flat space-time, R = 0 and the equations become free field equations. In curved space-time the curvature scalar term becomes a negative mass counter term reminiscent of the corresponding negative term in the Wheeler-DeWitt equation.

Another possible prototype lagrangian

$$\mathcal{L} = (\varphi_1 + \varphi_2)^2 \sqrt{g}\, R/2 + a\bar{\psi}(i\gamma^\mu\partial/\partial x^\mu - m)\psi + \partial\varphi_1/\partial x_\mu \partial\varphi_2/\partial x^\mu - m^2\, \varphi_1\varphi_2 \tag{7.28}$$

leads to an interaction between the pseudoQuantum φ_2 field and gravitation:

$$\Box\varphi_2 + m^2\varphi_2 - (\varphi_1 + \varphi_2)\sqrt{g}\, R = 0 \tag{7.29}$$
$$\Box\varphi_1 - m^2\varphi_1 = 0$$

7.3 The Eight Coupling Constants and their Eight PseudoQuantum Field Vacuum Expectation Values

As we mentioned in section 7.1 our Grand Unified Theory of Everything (GUTE) has eight coupling constants:

- The Strong interaction coupling constant field g_S.
- The ElectroWeak SU(2) coupling constant g_{EW}.
- The ElectroWeak U(1) coupling constant g'_{EW}.
- The Dark ElectroWeak SU(2) coupling constant g_{EWD}.
- The Dark ElectroWeak U(1) coupling constant g'_{EWD}.
- The Layer Group U(4) coupling constant[20] g_V.
- The Generation gauge field U(4) coupling constant g_G.
- The complex gravitational coupling constant $g_{CG} = \kappa^{-1} = (4\pi G)^{-1/2}$.

Based on the discussions of the previous sections we can define pseudoQuantum fields for these couplings by

- The strong interaction coupling constant vacuum expectation value $\Phi_1 = m_1 g_S$.
- The ElectroWeak SU(2) coupling constant vacuum expectation value $\Phi_2 = m_2 g_{EW}$.
- The ElectroWeak U(1) coupling constant vacuum expectation value $\Phi_3 = m_3 g'_{EW}$.
- The Dark ElectroWeak SU(2) coupling constant vacuum expectation value $\Phi_4 = m_4 g_{EWD}$.
- The Dark ElectroWeak U(1) coupling constant vacuum expectation value $\Phi_5 = m_5 g'_{EWD}$
- The Layer Group U(4) coupling constant vacuum expectation value $\Phi_6 = m_6 g_V$.
- The Generation gauge field U(4) coupling constant vacuum expectation value $\Phi_7 = m_7 g_G$.
- The gravitational coupling constant vacuum expectation value $\Phi_8 = g_{CG} = \kappa^{-1} = (4\pi G)^{-1/2}$.

The seven masses, m_1, m_2, … , m_7 may be equal or they may have different values. *It is also possible that all masses may be equal to κ^{-1}, which would yield*

- The strong interaction coupling constant vacuum expectation value $\Phi_1 = \kappa^{-1} g_S$.
- The ElectroWeak SU(2) coupling constant vacuum expectation value $\Phi_2 = \kappa^{-1} g_{EW}$.
- The ElectroWeak U(1) coupling constant vacuum expectation value $\Phi_3 = \kappa^{-1} g'_{EW}$.
- The Dark ElectroWeak SU(2) coupling constant vacuum expectation value $\Phi_4 = \kappa^{-1} g_{EWD}$.
- The Dark ElectroWeak U(1) coupling constant vacuum expectation value $\Phi_5 = \kappa^{-1} g'_{EWD}$.

---

[20] This coupling constant appears in Blaha (2016a).

- The Layer Group U(4) coupling constant vacuum expectation value $\Phi_6 = \kappa^{-1} g_V$.
- The Generation gauge field U(4) coupling constant vacuum expectation value $\Phi_7 = \kappa^{-1} g_G$.
- The gravitational coupling constant vacuum expectation value $\Phi_8 = g_{CG} = \kappa^{-1} = (4\pi G)^{-½}$.

Then scaling the above vacuum expectation values by κ^{-1} would give:[21]

- The strong interaction coupling constant[22] vacuum expectation value $\Phi_1' = g_S = 1.22$
- The ElectroWeak SU(2) coupling constant vacuum expectation value $\Phi_2' = g_{EW} = 0.619$.
- The ElectroWeak U(1) coupling constant vacuum expectation value $\Phi_3' = g'_{EW} = 0.347$.
- The Dark ElectroWeak SU(2) coupling constant vacuum expectation value $\Phi_4' = g_{EWD}$. (7.30)
- The Dark ElectroWeak U(1) coupling constant vacuum expectation value $\Phi_5' = g'_{EWD}$.
- The Layer Group U(4) coupling vacuum expectation value $\Phi_6 = g_V$.
- The Generation gauge field U(4) coupling constant vacuum expectation value $\Phi_7' = g_G$.
- The gravitational coupling constant vacuum expectation value $\Phi_8' = 1$.

The *scaled* (known) vacuum expectation values,[23] which are mostly in fact the coupling constants, have a comparable range of values[24] as opposed to the range of values for the unscaled constants which range from the ultra-small gravitational vacuum expectation value to values, perhaps, within a few orders of magnitude of unity.

Given the range of known values above, it appears reasonable to conjecture that the unknown values would also be of the order of unity. The known coupling constant values in eq. 7.30 are of comparable value, which suggests that our Theory of Everything, at current energies, may be close to the GUT level at which coupling constants are equal.

End of Extract

We thus see that in the pseudoquantum formalism for the vacuum expectation values of coupling constants $\Phi_1' = g_S = 1.22$ while the gravitation equivalent $\Phi_8' = 1$ is almost the same. We conclude that uniting the Strong Interaction and Gravitation is not unreasonable.

[21] All coupling constant values are based on data extracted from K. A. Olive et al (Particle Data Group), Chinese Physics **C38**, 090001 (2014).

[22] Based on the running coupling constant value $\alpha_s (M_Z^2) = 0.1193 \pm 0.0016$.

[23] The closeness of all the values to one is suggestive: The value $\alpha = 1$ (or $e = (4\pi)^{½} = 3.54$) was the value found in our calculation in the Johnson, Baker, Willey model of QED (Appendix A). Perhaps a larger calculation along the lines of our paper in massless ElectroWeak theory might yield scaled coupling constant values near unity.

[24] The weakness of the ElectroWeak interactions is primarily due to the large masses of the Z and W vector bosons – not the values of their coupling constants g and g'.

4. New Formulation of Complex General Relativity and the Strong Interaction

This chapter describes a unified formalism for Complex (and real-valued) General Relativity and the Strong Interaction.[25] It is largely based on a 1976 essay by the author (appendix D), a 1975 paper by the author, and Blaha (2016c) and (2015a) as well as earlier books.

4.1 Vierbein (Tetrad) Form of General Relativity

Weyl[26] first introduced the *vierbein* form of General Relativity in order to accommodate fermion spin within General Relativity. Since then a number of papers[27,28] have appeared on the vierbein formalism. The basic relations of the vierbein formalism are:

1. A vierbein (tetrad) has the form[29] $l^{\mu a}(x)$ where μ is a General Relativistic vector index and a is a Special Relativistic index for a point in a 4-dimensional space-time tangent space to a curved space-time. Both indices range from 0 to 3.

2. The curve space-time metric tensor is defined by

$$g_{\mu\nu} = \eta_{ab} l_{\mu}{}^{a}(x) l_{\nu}{}^{b}(x) \tag{4.1}$$

where $\eta_{ab} = \text{diag}(1, -1, -1, -1)$ is the flat space-time, tangent space metric.

3. It is convenient to establish a matrix form of a vierbein with

---

[25] This unified theory of Gravity and the Strong Interaction is unrelated to Kaluza-Klein theories (and derivatives thereof). The theories attempting to unify gravity and Electromagnetism had a different basis and typically relied on a curled fifth dimension. Our theory differs in origin and in detail. We espouse a 16-dimensional space without curling that constitutes what we have called the Megaverse. Kaluza-Klein theories are described in Kaluza, Theodor, "Zum Unitätsproblem in der Physik". Sitzungsber. Preuss. Akad. Wiss. *Berlin. (Math. Phys.)*: 966–972 (1921); Klein, Oskar, "Quantentheorie und fünfdimensionale Relativitätstheorie". Zeitschrift für Physik A **37** (12): 895–906 (1926; Klein, Oskar, "The Atomicity of Electricity as a Quantum Theory Law". *Nature* **118**: 516 (1926); and subsequent related papers such as Lisa Randall and Raman Sundrum, Phys. Rev. Lett., **83,** 3370 (1999).

[26] H. Weyl, Z. Physik. **56**, 330 (1929).

[27] R. Utiyama, Phys. Rev. **101**, 1597 (1956), T. W. B. Kibble, Jour. Math. Phys. **2**, 212 (1961), J. Schwinger, Phys. Rev. **130**, 800 (1963), J. Schwinger, Phys. Rev. **130**, 1253 (1963). **We generally use their notation in this chapter.**

[28] C. Isham, A. Salam, and J. Strathdee, Phys. Rev. D**3**, 867 (1971); ________,Lett. Al Nuov. Cim., **5**, 969 (1972) and references therein.

[29] Typographic note: Since Times Roman type does not distinguish between 1 (one) and l (el) the vierbein l is distinguished by indices or appearance in a matrix equation.

$$l^{\mu}(x) = l^{\mu a}(x)\gamma_a \qquad (4.2)$$

where the four matrices γ_a are the familiar Dirac matrices.

4. Under a local Special Relativistic transformation S a spin ½ field transforms as

$$\psi(x) \rightarrow S(x)\psi(x) \qquad (4.3)$$

and the matrix form of a vierbein transforms as

$$l^{\mu}(x) \rightarrow S(x)l^{\mu}(x)S^{-1}(x) \qquad (4.4)$$

5. Under a General Relativistic transformation a vierbein transforms as

$$l'^{\mu a}(x') = \partial x_{\nu}/\partial x'_{\mu}\ l^{\nu a}(x) \qquad (4.5)$$

Other features of the vierbein formalism can be found in papers in the aforementioned footnotes and in Weinberg (1972) as well as other books.

We will assume that the x coordinate system is real-valued initially. Then we will map it to complex values in keeping with Blaha (2015a) and our other books. We showed many years ago that faster-than-light motion is possible and that it requires complex-valued coordinates and complex Lorentz transformations.

4.2 Extension of the Vierbein Formalism to Incorporate the Strong Interaction

We can simply extend[30] the above General Relativistic vierbein formalism to include the the SU(3) Strong Interaction by adding an additional SU(3) index to the vierbein. Thus the extended formalism includes

1. An SU(3) extended vierbein (tetrad) has the form[31] $l^{\mu a i}(x)$ where the additional index i for SU(3) ranges from 1 through 8.

2. The curve space-time metric tensor now is defined by

$$g_{\mu\nu} = \eta_{ab} l_{\mu}{}^{ai}(x) l_{\nu}{}^{b}{}_{i}(x) \qquad (4.6)$$

with a sum over i.

[30] This extension, with a slight modification, appears in the author's Gravity Research Foundation 1976 Essay competition submission for which he received Honorable Mention. It is reproduced in appendix D.

[31] Typographic note: Since Times Roman type does not distinguish between 1 (one) and l (el) the vierbein l is distinguished by indices or appearance in a matrix equation.

3. It is again convenient to establish a matrix form of a vierbein with

$$l^{\mu}(x) = l^{\mu a i}(x)\gamma_a T_i \qquad (4.7)$$

where the eight matrices T_i are SU(3) generators in the fundamental 3 representation.

4. Under an SU(3) local gauge transformation C the *unified* vierbein, in matrix form, transforms as

$$l^{\mu}(x) \rightarrow C(x)l^{\mu}(x)C^{-1}(x) \qquad (4.8)$$

Transformations under General Relativity and Special Relativity have the same form as above.

4.3 Physical Interpretation of the SU(3) Extended Vierbein

The extended vierbein $l^{\mu a i}(x)$ can be viewed as located at a point in a 32-dimensional real-valued space.

$$l^{\mu a i}(x) = (\partial \xi_X{}^{ai}(x)/\partial x_{\mu})_{X=h(x)} \qquad (4.9)$$

where $\xi_X{}^{ai}$ is a set of locally inertial coordinates located at a 32-dimensional point X, and x = h(x) is a 4-dimensional point in a tangent subspace of the 32-dimensional space:

$$X = h(x) \qquad (4.10)$$

The relation between real 4-dimensional coordinates x and the 32-dimensional coordinates X is an embedding of a 4-dimensional surface within a 32-dimensional real space when account is taken of the range of possible x values. We have considered such embeddings in Blaha (2015a), and in earlier books, and developed a theory of a 16-dimensional complex-valued space (the *Megaverse*) that contains our universe and probably many other universes. The study, in which we are now engaged, adds to the reasons given in earlier books for belief in the Megaverse.[32]

4.4 The Megaverse – The 16-dimensional Space with Complex-valued Coordinates

The 32-dimensional flat, real-valued, tangent space considered above can be mapped onto a flat 16-dimensional universe with complex-valued coordinates in a straightforward way. Upon making that transition we can turn the above process around and view our curved 4-dimensional universe as a flat complex-valued subspace residing within a 16-dimensional,

[32] Many years ago, the term Megaverse was introduced by William James. It signifies a space with many resident universes. The word multiverse was popularized by the work of Everritt and signifies many parallel universes based on quantum mechanical considerations.

complex-valued, tangent space now grown to a full space by extension to all possible points X.[33] This is the Megaverse.

Since the coordinates of our space-time are necessarily complex-valued (Blaha (2015a) and earlier books), it is natural for the Megaverse to have complex coordinates as well.

The Megaverse was considered flat in the previous section. Now we can introduce interactions (and a definition of mass-energy) that make the Megaverse curved as well.[34]

Thus we view the Megaverse as a curved, complex-valued, 16-dimensional space containing universes interacting with each other through long-range forces such as the elusive Baryon Number force and the Dark Baryon Number force. The universes then appear to be strewn through the Megaverse like galaxies within our universe. Universes may collide and perhaps combine over incredibly long time frames. Blaha (2015a) and (2015b) considers universe dynamics in some detail and proposes a quantum field theory of universe particles, *uons*, based on the form of the Wheeler-DeWitt equation. The reader is referred to these works for further discussion. We return to the consideration of the unification of Gravitation and the Strong Interaction.

A real-valued, 4-dimensional curved universe can be mapped to a surface in a 10-dimensional Euclidean space since the metric tensor has 10 components. For a universe with 10 complex-valued metric tensor components a real-valued Euclidean space must have at least 20 dimensions or ten complex-valued dimensions. Since the Megaverse has 16 complex-valued dimensions our universe, as well as other 4-dimensional universes, can be mapped to it as surfaces using maps such as $\mathbf{X} = \mathbf{h}(\mathbf{x})$ where $\mathbf{X}$ is a 16-vector, $\mathbf{x}$ is a 4-vector and $\mathbf{h}()$ is a a 16-vector function.

4.5 Dynamics of Unification

This section describes the basic fields and their dynamics.

4.5.1 The Gravitation and Strong Interaction Fields

The spinor connection used in formulations of vierbein gravity is $B^1{}_{\mu ab}(x)$ where a and b are tangent space indices.[35] The vector is combined with γ matrices for use in matrix equations:

$$B^{1\mu} = B^{1\mu}{}_{ab}\Sigma^{ab} \tag{4.11}$$

where

$$\Sigma^{ab} = i\,[\gamma^a, \gamma^b]/4 \tag{4.12}$$

---

[33] An informed reader may ask: If the Strong Interactions are confined as they appear to be, how can you define a space which, in part, has an SU(3) Strong Interaction part? Would not the space be 'confined' in some way? The answer lies in realizing that coordinates are not confined. Particles are confined. Coordinates are not fields, and the metric and the curvature tensor of the space are independent of the Strong Interaction outside of regions containing quarks and gluons. Within a hadron or quark-gluon plasma, the interplay of Gravitation and the Strong Interaction will have effects on the curvature of space.

[34] See Blaha (2015a) and (2015b).

[35] We introduce superscripts '1' and '2' on vectors A^μ and B^μ associated with the affine connections decribed in this section and following sections.

Under a local Lorentz transformation S

$$B^{1\mu}(x) \rightarrow S(x)B^{1\mu}(x)S^{-1}(x) - i\, S(x)\partial^{\mu}S^{-1}(x) \tag{4.13}$$

Similarly a spin ½ field transforms as

$$(\partial^{\mu} + i\, B^{1\mu})\psi \rightarrow S(\partial^{\mu} + i\, B^{1\mu})\psi \tag{4.14}$$

The SU(3) gauge fields are defined as $A^{1i\mu}(x)$ $A^{2i\mu}(x)$ and for i = 1, … , 8. Using SU(3) generators we define the matrix form by $A^{\mu}(x) = A^{i\mu}(x)T_i$. Under an SU(3) gauge transformation C the gauge field transforms as

$$A^{1\mu}(x) \rightarrow C(x)A^{1\mu}(x)C^{-1}(x) - i\, C(x)\partial^{\mu}C^{-1}(x) \tag{4.15a}$$

And

$$A^{2\mu}(x) \rightarrow C(x)A^{2\mu}(x)C^{-1}(x) \tag{4.15b}$$

4.5.2 The Gravitation Affine Connection

The affine connection is most often viewed as a derived quantity—part of the derivation of the curvature tensor in General Relativity. It is typically derived from manipulations of the metric $g_{\mu\nu}$. However, the affine connection can also be viewed as a set of independent fields that become related to the metric via dynamic equations.

Some years ago A. Einstein and H. Weyl[36] pointed out that the metric and the affine connection should be treated as independent quantities and subject to independent arbitrary infinitesimal variations:

> "In contrast to Einstein's original "metric" conception in terms of the $g_{\nu\mu}$ there was later developed, by Eddington, by Einstein himself, and recently by Schrödinger, an affine field theory operating with the components $\Gamma^{\sigma}{}_{\nu\mu}$ of an affine connection. But in 1925 Einstein also advocated a "mixed" formulation by means of a lagrangian in which both the $g_{\nu\mu}$ and the $\Gamma^{\sigma}{}_{\nu\mu}$ are taken as basic field quantities and submitted to independent arbitrary infinitesimal variations.[37] In certain respects this seems to be the most natural procedure."

Following this approach we introduce affine connections. For the gravitational part of the vierbein we take the spinor affine connection to be given in eq. 4.11.

Further we define another affine connection, which is related to the usual General Relativistic gravitational connection $\Gamma^{\sigma}{}_{\nu\mu}$:

$$B'^{\sigma}{}_{\nu\mu} = B'^{\sigma ab}{}_{\nu\mu}\Sigma_{ab} \tag{4.16}$$

with Σ_{ab} defined by eq. 4.12. $B'^{\sigma}{}_{\nu\mu}$ will be determined by the dynamic equations.

---

[36] H. Weyl, Phys. Rev. **77**, 699 (1950).

[37] A. Einstein, Sitzungsber., Preuss. Akad. Der Wissensch. (1925), p. 414.

Under a local Lorentz transformation S(x) we require both $B'^{\sigma}{}_{\nu\mu}$ and $B'^{\sigma ab}{}_{\nu\mu}$ transform homogeneously

$$B'^{\sigma ab}{}_{\nu\mu} \rightarrow S(x)B'^{\sigma ab}{}_{\nu\mu}S^{-1}(x) \tag{4.17}$$
$$B'^{\sigma}{}_{\nu\mu} \rightarrow S(x)B'^{\sigma}{}_{\nu\mu}S^{-1}(x) \tag{4.18}$$

We futher define $B'^{\sigma}{}_{\nu\mu}$ as

$$B'^{\alpha ab}{}_{\nu\mu}\, l_{\alpha a}{}^{i}(x) l^{\sigma}{}_{bi}(x) = i\, \Gamma^{\sigma}{}_{\nu\mu}/2 \tag{4.19}$$

with an implicit sum over i.

Thus

$$B'^{\sigma}{}_{\nu\mu}\times l_{\sigma} \equiv [B'^{\sigma}{}_{\nu\mu}, l_{\sigma}] = \Gamma^{\sigma}{}_{\nu\mu} l_{\sigma} \tag{4.20}$$

for all implicit SU(3) indices i.

4.5.3 The SU(3) Affine Connection

We now define an additional SU(3) Yang-Mills affine connection (a general coordinate transformation tensor) as:

$$A'^{\sigma}{}_{\nu\mu} = A'^{\sigma i}{}_{\nu\mu} T_i \tag{4.21}$$

Under a local gauge transformation C(x) we define its gauge transformation to be homogeneous:

$$A'^{\sigma}{}_{\nu\mu}(x) \rightarrow C(x)A'^{\sigma}{}_{\nu\mu}(x)C^{-1}(x) \tag{4.22}$$

Based on the above stated views of Einstein and Weyl that the metric and the affine connection should be treated as independent quantities and subject to independent arbitrary infinitesimal variations we define the SU(3) affine connection in terms of a new SU(3) gauge field A'_{μ} as

$$A'^{\sigma}{}_{\nu\mu} = g^{\sigma}{}_{\nu} A^{2}{}_{\mu} + g^{\sigma}{}_{\mu} A^{2}{}_{\nu} \tag{4.23}$$

where

$$A^{2}{}_{\mu} = A^{2i}{}_{\mu} T_i \tag{4.24}$$

Eq. 4.23 maintains the usual affine connection symmetry in ν and μ of $A'^{\sigma}{}_{\nu\mu}$. Eq. 4.22 implies that A'_{μ} is a Lorentz vector that transforms homogeneously under local SU(3) gauge transformations C(x).

$$A^{2\mu}(x) \rightarrow C(x)A^{2\mu}(x)C^{-1}(x) \tag{4.25}$$

4.5.4 Covariant Derivatives and the Unified Curvature Tensor

In this subsection we define the covariant derivative, curvature tensor, Ricci tensor, and curvature scalar for the unification of gravity and SU(3) in a 32-dimensional real space without the use of the subsidiary affine connections defined by eqs. 4.16 and 4.22.

Based on the preceding sections we use the following covariant derivative of a vector in 32-dimensional space:[38]

---

[38] We use the superscript '1' to distinguish from secondary connections introduced in the following sections,

$$\begin{aligned} D_\nu V_\mu &= (\partial_\nu + iB^1{}_\nu + iA^1{}_\nu)V_\mu - \Gamma^\sigma{}_{\nu\mu}V_\sigma \\ &= (\partial_\nu + iC_\nu)V_\mu - \Gamma^\sigma{}_{\nu\mu}V_\sigma \\ &= [g^\sigma{}_\mu\partial_\nu + ig^\sigma{}_\mu C_\nu - \Gamma^\sigma{}_{\nu\mu}]V_\sigma \\ &= [g^\sigma{}_\mu\partial_\nu + iD^\sigma{}_{\mu\nu}]V_\sigma \end{aligned} \tag{4.26}$$

using eqs. 4.19 and 4.23 where

$$C_\mu = B^1{}_\mu + A^1{}_\mu \tag{4.27}$$
$$D^\sigma{}_{\mu\nu} = g^\sigma{}_\mu C_\nu + i\Gamma^\sigma{}_{\nu\mu} \tag{4.28}$$

Eq. 4.26 enables us to simply derive the Riemann-Christoffel curvature tensor, and then its contractions $R_{\mu\nu}$ and R using

$$(D_\nu D_\mu - D_\mu D_\nu)V_\sigma = R^\beta{}_{\sigma\nu\mu}V_\beta \tag{4.29}$$

where

$$V_\sigma = V_\sigma{}^{ai}(x)\gamma_a T_i \tag{4.30}$$

Then

$$\begin{aligned} D_\nu D_\mu V_\sigma = \{g^\alpha{}_\mu(\partial_\nu + iB^1{}_\nu + iA^1{}_\nu) - \Gamma^\alpha{}_{\mu\nu}\}\{g^\beta{}_\sigma(\partial_\alpha + iB^1{}_\alpha + iA^1{}_\alpha)V_\beta - \Gamma^\beta{}_{\sigma\alpha}V_\beta\} - \\ - \Gamma^\gamma{}_{\nu\sigma}\{g^\alpha{}_\gamma(\partial_\mu + iB^1{}_\mu + iA^1{}_\mu)V_\alpha - \Gamma^\alpha{}_{\gamma\mu}V_\alpha\} \end{aligned} \tag{4.31}$$

and

$$\begin{aligned} R^\beta{}_{\sigma\nu\mu}V_\beta &= g^\alpha{}_\mu(\partial_\nu + iB^1{}_\nu + iA^1{}_\nu)g^\beta{}_\sigma(\partial_\alpha + iB^1{}_\alpha + iA^1{}_\alpha)V_\beta - \Gamma^\alpha{}_{\mu\nu}g^\beta{}_\sigma(\partial_\alpha + iB^1{}_\alpha + iA^1{}_\alpha)V_\beta + \\ &+ \Gamma^\alpha{}_{\mu\nu}\Gamma^\beta{}_{\sigma\alpha}V_\beta - g^\alpha{}_\mu(\partial_\nu + iB^1{}_\nu + iA^1{}_\nu)\Gamma^\beta{}_{\sigma\alpha}V_\beta - \Gamma^\gamma{}_{\nu\sigma}\{g^\alpha{}_\gamma(\partial_\mu + iB^1{}_\mu + iA^1{}_\mu)V_\alpha - \Gamma^\alpha{}_{\gamma\mu}V_\alpha\} - \\ &- \{\mu\leftrightarrow\nu\} \\ &= ig^\beta{}_\sigma F^1{}_{\nu\mu}V_\beta + (ig^\beta{}_\sigma B^1{}_{\nu\mu} + \partial_\mu\Gamma^\beta{}_{\sigma\nu} - \partial_\nu\Gamma^\beta{}_{\sigma\mu} + \Gamma^\gamma{}_{\nu\sigma}\Gamma^\beta{}_{\gamma\mu} - \Gamma^\gamma{}_{\mu\sigma}\Gamma^\beta{}_{\gamma\nu})V_\beta \\ &= R_{SU(3)}{}^\beta{}_{\sigma\nu\mu}V_\beta + R^\beta{}_{\sigma\nu\mu}V_\beta \end{aligned} \tag{4.32}$$

where

$$\begin{aligned} F^1{}_{\kappa\mu} &= \partial A^1{}_\mu/\partial x^\kappa - \partial A^1{}_\kappa/\partial x^\mu + i[A^1{}_\kappa \times A^1{}_\mu] \\ B^1{}_{\kappa\mu} &= \partial B^1{}_\mu/\partial x^\kappa - \partial B^1{}_\kappa/\partial x^\mu + iB^1{}_\kappa \times B^1{}_\mu - iB^1{}_\mu \times B^1{}_\kappa \end{aligned} \tag{4.33}$$

Thus the total SU(3) and Gravity Curvature tensor is

$$\begin{aligned} R_{tot}{}^\beta{}_{\sigma\nu\mu} &= ig^\beta{}_\sigma F^1{}_{\nu\mu} + (ig^\beta{}_\sigma B^1{}_{\nu\mu} + \partial_\mu\Gamma^\beta{}_{\sigma\nu} - \partial_\nu\Gamma^\beta{}_{\sigma\mu} + \Gamma^\gamma{}_{\nu\sigma}\Gamma^\beta{}_{\gamma\mu} - \Gamma^\gamma{}_{\mu\sigma}\Gamma^\beta{}_{\gamma\nu}) \\ &= R_{SU(3)}{}^\beta{}_{\sigma\nu\mu} + R^\beta{}_{\sigma\nu\mu} \end{aligned} \tag{4.34}$$

Note $R_{tot}{}^\beta{}_{\sigma\nu\mu}$ factorizes into an SU(3) part and a Riemann-Christoffel curvature tensor part:

$$R_{SU(3)}{}^\beta{}_{\sigma\nu\mu} = ig^\beta{}_\sigma F^1{}_{\nu\mu} \tag{4.35}$$

$$R^\beta{}_{\sigma\nu\mu} = ig^\beta{}_\sigma B^1{}_{\nu\mu} + \partial_\mu\Gamma^\beta{}_{\sigma\nu} - \partial_\nu\Gamma^\beta{}_{\sigma\mu} + \Gamma^\gamma{}_{\nu\sigma}\Gamma^\beta{}_{\gamma\mu} - \Gamma^\gamma{}_{\mu\sigma}\Gamma^\beta{}_{\gamma\nu} \tag{4.36}$$

$R_{tot}{}^{\beta}{}_{\sigma\nu\mu}$ *is the Riemann-Christoffel curvature tensor for the real 32-dimensional space generated by SU(3) and 4-dimensional General Coordinate transformations.* Note also the anti-symmetry in μ and ν.

The total Ricci tensor is

$$\begin{aligned} R_{tot\sigma\mu} &= R_{tot}{}^{\beta}{}_{\sigma\beta\mu} \\ &= iF^{1}{}_{\sigma\mu} + (iB^{1}{}_{\sigma\mu} + \partial_{\mu}\Gamma^{\beta}{}_{\sigma\beta} - \partial_{\beta}\Gamma^{\beta}{}_{\sigma\mu} + \Gamma^{\gamma}{}_{\beta\sigma}\Gamma^{\beta}{}_{\gamma\mu} - \Gamma^{\gamma}{}_{\mu\sigma}\Gamma^{\beta}{}_{\gamma\beta}) \\ &= R_{SU(3)\sigma\mu} + R_{\sigma\mu} \end{aligned} \tag{4.37}$$

The curvature scalar is

$$R_{tot} = g^{\sigma\mu}R_{tot\sigma\mu} = g^{\sigma\mu}(\partial_{\mu}\Gamma^{\beta}{}_{\sigma\beta} - \partial_{\beta}\Gamma^{\beta}{}_{\sigma\mu} + \Gamma^{\gamma}{}_{\beta\sigma}\Gamma^{\beta}{}_{\gamma\mu} - \Gamma^{\gamma}{}_{\mu\sigma}\Gamma^{\beta}{}_{\gamma\beta}) \tag{4.38}$$

Eq. 4.37 is the Ricci tensor for the real 32-dimensional space generated by SU(3) and 4-dimensional General Coordinate transformations.Eq. 4.38 is its curvature scalar. Note the curvature scalar is independent of SU(3) and the spinor connection $B^{1}{}_{\mu}$.

4.5.5 Covariant Derivatives and the Unified Curvature Tensor with the Additional SU(3) and Vierbein Affine Connections

In this subsection we define the covariant derivative, curvature tensor, Ricci tensor, and curvature scalar for the unification of gravity and SU(3) in a 32-dimensional real space WITH the use of the subsidiary SU(3) affine connection defined by eq. 4.23, and a secondary metric, spinor connetion, and gravitational affine connection. In accord with Einstein and Weyl we will subject these secondary quantities to independent variations when deriving the field equations.

The motivation for these secondary fields is to enable us to use the canonical lagrangian formalism for higher derivative – 4th order – SU(3) and gravitation theories. *These secondary quantities, combined with those of the previous subsection, will enable us to define a canonical lagrangian formulation of gravity unified with SU(3) with a modified form of the gravitational potential, and a SU(3) interaction with an r potential yielding quark confinement.*

This subsection will purposefully begin, analogously, like the previous subsection. We begin by defining subsidiary quantities to support a higher derivative, lagrangian formulation:

$$g^{2\mu\nu} = g^{2\nu\mu} \tag{4.39}$$

$$B^{2\mu} = B^{2\mu}{}_{ab}\Sigma^{ab} \tag{4.40}$$

where Σ^{ab} is defined by eq. 4.12, and a secondary affine connection:

$$\Gamma^{2\lambda}{}_{\mu\nu} = \tfrac{1}{2}g^{2\lambda\alpha}(\partial_{\mu}g^{2}{}_{\alpha\nu} + \partial_{\nu}g^{2}{}_{\alpha\mu} - \partial_{\alpha}g^{2}{}_{\mu\nu}) \tag{4.41}$$

We use the following generalized covariant derivative of a vector in our 32-dimensional space:[39]

$$D_\nu V_\mu = (\partial_\nu + iF_\nu)V_\mu - H^\sigma{}_{\nu\mu}V_\sigma \tag{4.42}$$
$$= [g^\sigma{}_\mu\partial_\nu + ig^\sigma{}_\mu F_\nu - H^\sigma{}_{\nu\mu}]V_\sigma$$
$$= [g^\sigma{}_\mu\partial_\nu + iD^\sigma{}_{\mu\nu}]V_\sigma$$

using eqs. 4.19 and 4.23 where

$$F_\mu = B^1{}_\mu + A^1{}_\mu + B^2{}_\mu + A^2{}_\mu \tag{4.43}$$
$$H^\sigma{}_{\nu\mu} = \Gamma^\sigma{}_{\nu\mu} + \Gamma^{2\sigma}{}_{\nu\mu} \tag{4.43a}$$
$$D^\sigma{}_{\mu\nu} = g^\sigma{}_\mu F_\nu + iH^\sigma{}_{\nu\mu} \tag{4.44}$$

Eq. 4.42 enables us to derive the Riemann-Christoffel curvature tensor, and then its contractions $R_{\mu\nu}$ and R using

$$(D_\nu D_\mu - D_\mu D_\nu)V_\sigma = R^\beta{}_{\sigma\nu\mu}V_\beta \tag{4.45}$$

where

$$V_\sigma = V_\sigma{}^{ai}(x)\gamma_a T_i \tag{4.46}$$

Then

$$D_\nu D_\mu V_\sigma = \{g^\alpha{}_\mu(\partial_\nu + iF_\nu) - H^\alpha{}_{\mu\nu}\}\{g^\beta{}_\sigma(\partial_\alpha + iF_\alpha)V_\beta - H^\beta{}_{\sigma\alpha}V_\beta\} - H^\gamma{}_{\nu\sigma}\{g^\alpha{}_\gamma(\partial_\mu + iF_\mu)V_\alpha - H^\alpha{}_{\gamma\mu}V_\alpha\} \tag{4.47}$$

As a result

$$R'^\beta{}_{\sigma\nu\mu}V_\beta = g^\alpha{}_\mu(\partial_\nu + iF_\nu)g^\beta{}_\sigma(\partial_\alpha + iF_\alpha)V_\beta - H^\alpha{}_{\mu\nu}g^\beta{}_\sigma(\partial_\alpha + iF_\alpha)V_\beta +$$
$$+ H^\alpha{}_{\mu\nu}H^\beta{}_{\sigma\alpha}V_\beta - g^\alpha{}_\mu(\partial_\nu + iF_\nu)H^\beta{}_{\sigma\alpha}V_\beta - H^\gamma{}_{\nu\sigma}\{g^\alpha{}_\gamma(\partial_\mu + iF_\mu)V_\alpha - H^\alpha{}_{\gamma\mu}V_\alpha\} -$$
$$- \{\mu\leftrightarrow\nu\}$$

$$= ig^\beta{}_\sigma(\partial_\nu F_\mu - \partial_\mu F_\nu - i[F_\nu, F_\mu])V_\beta + (\partial_\mu H^\beta{}_{\sigma\nu} - \partial_\nu H^\beta{}_{\sigma\mu} + H^\gamma{}_{\nu\sigma}H^\beta{}_{\gamma\mu} - H^\gamma{}_{\mu\sigma}H^\beta{}_{\gamma\nu})V_\beta$$

$$= ig^\beta{}_\sigma(F^1{}_{\nu\mu} + F^2{}_{\nu\mu})V_\beta + (ig^\beta{}_\sigma B^1{}_{\nu\mu} + ig^\beta{}_\sigma B^2{}_{\nu\mu} + \partial_\mu H^\beta{}_{\sigma\nu} - \partial_\nu H^\beta{}_{\sigma\mu} + H^\gamma{}_{\nu\sigma}H^\beta{}_{\gamma\mu} - H^\gamma{}_{\mu\sigma}H^\beta{}_{\gamma\nu})V_\beta$$

$$= R'_{SU(3)}{}^\beta{}_{\sigma\nu\mu}V_\beta + R'_G{}^\beta{}_{\sigma\nu\mu}V_\beta \tag{4.48}$$

where

$$R'_{SU(3)}{}^\beta{}_{\sigma\nu\mu} = ig^\beta{}_\sigma(F^1{}_{\nu\mu} + F^2{}_{\nu\mu}) \tag{4.49}$$

$$R'_G{}^\beta{}_{\sigma\nu\mu} = ig^\beta{}_\sigma(B^1{}_{\nu\mu} + B^2{}_{\nu\mu}) + \partial_\mu\Gamma^\beta{}_{\sigma\nu} - \partial_\nu\Gamma^\beta{}_{\sigma\mu} + \Gamma^\gamma{}_{\nu\sigma}\Gamma^\beta{}_{\gamma\mu} - \Gamma^\gamma{}_{\mu\sigma}\Gamma^\beta{}_{\gamma\nu} + \partial_\mu\Gamma^{2\beta}{}_{\sigma\nu} - \partial_\nu\Gamma^{2\beta}{}_{\sigma\mu} +$$
$$+\Gamma^{2\gamma}{}_{\nu\sigma}\Gamma^{2\beta}{}_{\gamma\mu} - \Gamma^{2\gamma}{}_{\mu\sigma}\Gamma^{2\beta}{}_{\gamma\nu} + \Gamma^\gamma{}_{\nu\sigma}\Gamma^{2\beta}{}_{\gamma\mu} - \Gamma^\gamma{}_{\mu\sigma}\Gamma^{2\beta}{}_{\gamma\nu} + \Gamma^{2\gamma}{}_{\nu\sigma}\Gamma^\beta{}_{\gamma\mu} - \Gamma^{2\gamma}{}_{\mu\sigma}\Gamma^\beta{}_{\gamma\nu} \tag{4.50}$$
$$= ig^\beta{}_\sigma(B^1{}_{\nu\mu} + B^2{}_{\nu\mu}) + R^{1\beta}{}_{\sigma\nu\mu} + R^{2\beta}{}_{\sigma\nu\mu}$$

with

$$H^\beta{}_{\sigma\nu\mu} = \partial_\mu H^\beta{}_{\sigma\nu} - \partial_\nu H^\beta{}_{\sigma\mu} + H^\gamma{}_{\nu\sigma}H^\beta{}_{\gamma\mu} - H^\gamma{}_{\mu\sigma}H^\beta{}_{\gamma\nu} \tag{4.50a}$$
$$R^{1\beta}{}_{\sigma\nu\mu} = \partial_\mu\Gamma^\beta{}_{\sigma\nu} - \partial_\nu\Gamma^\beta{}_{\sigma\mu} + \Gamma^\gamma{}_{\nu\sigma}\Gamma^\beta{}_{\gamma\mu} - \Gamma^\gamma{}_{\mu\sigma}\Gamma^\beta{}_{\gamma\nu} \tag{4.51}$$

---

[39] We use the superscript '1' to distinguish primary connections from secondary connections labeled '2'.

$$R^{2\beta}{}_{\sigma\nu\mu\rho} = \partial_\mu\Gamma^{2\beta}{}_{\sigma\nu} - \partial_\nu\Gamma^{2\beta}{}_{\sigma\mu} + \Gamma^{2\gamma}{}_{\nu\sigma}\Gamma^{2\beta}{}_{\gamma\mu} - \Gamma^{2\gamma}{}_{\mu\sigma}\Gamma^{2\beta}{}_{\gamma\nu} + + \Gamma^{\gamma}{}_{\nu\sigma}\Gamma^{2\beta}{}_{\gamma\mu} - \Gamma^{\gamma}{}_{\mu\sigma}\Gamma^{2\beta}{}_{\gamma\nu} + \Gamma^{2\gamma}{}_{\nu\sigma}\Gamma^{\beta}{}_{\gamma\mu} - \Gamma^{2\gamma}{}_{\mu\sigma}\Gamma^{\beta}{}_{\gamma\nu} \quad (4.52)$$

and where

$$F^1{}_{\kappa\mu} = \partial A^1{}_\mu/\partial x^\kappa - \partial A^1{}_\kappa/\partial x^\mu + i[A^1{}_\kappa, A^1{}_\mu] \quad (4.53)$$
$$B^1{}_{\kappa\mu} = \partial B^1{}_\mu/\partial x^\kappa - \partial B^1{}_\kappa/\partial x^\mu + i[B^1{}_\kappa, B^1{}_\mu]$$
$$F^2{}_{\kappa\mu} = \partial A^2{}_\mu/\partial x^\kappa - \partial A^2{}_\kappa/\partial x^\mu + i[A^2{}_\kappa, A^2{}_\mu] + i[A^1{}_\kappa, A^2{}_\mu] + i[A^2{}_\kappa, A^1{}_\mu]$$
$$B^2{}_{\kappa\mu} = \partial B^2{}_\mu/\partial x^\kappa - \partial B^2{}_\kappa/\partial x^\mu + i[B^2{}_\mu, B^2{}_\kappa] + i[B^1{}_\mu, B^2{}_\kappa] + i[B^2{}_\mu, B^1{}_\kappa]$$

$R'^{\beta}{}_{\sigma\nu\mu}$ *is the Riemann-Christoffel curvature tensor for the real 32-dimensional space generated by SU(3) and 4-dimensional General Coordinate transformations including secondary connections and metric.*

Note $R'^{\beta}{}_{\sigma\nu\mu}$ factorizes into an SU(3) part and a Riemann-Christoffel curvature tensor part. For later use in defining a lagrangian we define

$$R'^{\beta}{}_{\sigma\nu\mu} = R'^1{}_{SU(3)}{}^{\beta}{}_{\sigma\nu\mu} + R'^2{}_{SU(3)}{}^{\beta}{}_{\sigma\nu\mu} + R'^1{}_G{}^{\beta}{}_{\sigma\nu\mu} + R'^2{}_G{}^{\beta}{}_{\sigma\nu\mu} \quad (4.54)$$

where

$$R'^1{}_{SU(3)}{}^{\beta}{}_{\sigma\nu\mu} = ig^{\beta}{}_{\sigma}F^1{}_{\nu\mu} \quad (4.55)$$
$$R'^2{}_{SU(3)}{}^{\beta}{}_{\sigma\nu\mu} = ig^{\beta}{}_{\sigma}F^2{}_{\nu\mu} \quad (4.56)$$
$$R'^1{}_G{}^{\beta}{}_{\sigma\nu\mu} = ig^{\beta}{}_{\sigma}B^1{}_{\nu\mu} + \partial_\mu\Gamma^{\beta}{}_{\sigma\nu} - \partial_\nu\Gamma^{\beta}{}_{\sigma\mu} + \Gamma^{\gamma}{}_{\nu\sigma}\Gamma^{\beta}{}_{\gamma\mu} - \Gamma^{\gamma}{}_{\mu\sigma}\Gamma^{\beta}{}_{\gamma\nu} \quad (4.57)$$
$$R'^2{}_G{}^{\beta}{}_{\sigma\nu\mu} = \partial_\mu\Gamma^{2\beta}{}_{\sigma\nu} - \partial_\nu\Gamma^{2\beta}{}_{\sigma\mu} + \Gamma^{2\gamma}{}_{\nu\sigma}\Gamma^{2\beta}{}_{\gamma\mu} - \Gamma^{2\gamma}{}_{\mu\sigma}\Gamma^{2\beta}{}_{\gamma\nu} + + \Gamma^{\gamma}{}_{\nu\sigma}\Gamma^{2\beta}{}_{\gamma\mu} - \Gamma^{\gamma}{}_{\mu\sigma}\Gamma^{2\beta}{}_{\gamma\nu} + \Gamma^{2\gamma}{}_{\nu\sigma}\Gamma^{\beta}{}_{\gamma\mu} - \Gamma^{2\gamma}{}_{\mu\sigma}\Gamma^{\beta}{}_{\gamma\nu} \quad (4.58)$$

The total Ricci tensor is

$$\begin{aligned} R'_{\sigma\mu} &= R'^{\beta}{}_{\sigma\beta\mu} \\ &= iF^1{}_{\sigma\mu} + iF^2{}_{\sigma\mu} + iB^1{}_{\sigma\mu} + iB^2{}_{\sigma\mu} + \partial_\mu\Gamma^{\beta}{}_{\sigma\beta} - \partial_\beta\Gamma^{\beta}{}_{\sigma\mu} + \Gamma^{\gamma}{}_{\beta\sigma}\Gamma^{\beta}{}_{\gamma\mu} - \Gamma^{\gamma}{}_{\mu\sigma}\Gamma^{\beta}{}_{\gamma\beta} + \\ &\quad + \partial_\mu\Gamma^{2\beta}{}_{\sigma\beta} - \partial_\beta\Gamma^{2\beta}{}_{\sigma\mu} + \Gamma^{2\gamma}{}_{\beta\sigma}\Gamma^{2\beta}{}_{\gamma\mu} - \Gamma^{2\gamma}{}_{\mu\sigma}\Gamma^{2\beta}{}_{\gamma\beta} + \Gamma^{\gamma}{}_{\beta\sigma}\Gamma^{2\beta}{}_{\gamma\mu} - \Gamma^{\gamma}{}_{\mu\sigma}\Gamma^{2\beta}{}_{\gamma\beta} + \Gamma^{2\gamma}{}_{\beta\sigma}\Gamma^{\beta}{}_{\gamma\mu} - \Gamma^{2\gamma}{}_{\mu\sigma}\Gamma^{\beta}{}_{\gamma\beta} \\ &= R'^1{}_{SU(3)}{}^{\beta}{}_{\sigma\beta\mu} + R'^2{}_{SU(3)}{}^{\beta}{}_{\sigma\beta\mu} + R'^1{}_G{}^{\beta}{}_{\sigma\beta\mu} + R'^2{}_G{}^{\beta}{}_{\sigma\beta\mu} \\ &= R'^1{}_{SU(3)\sigma\mu} + R'^2{}_{SU(3)\sigma\mu} + R'^1{}_{G\sigma\mu} + R'^2{}_{G\sigma\mu} = R'^1{}_{\sigma\mu} + R'^2{}_{\sigma\mu} \end{aligned} \quad (4.59)$$

Eq. 4.59 is the Ricci tensor for the real 32-dimensional space generated by SU(3) and 4-dimensional General Coordinate transformations when augmented with the secondary affine connections.

An additional Ricci-like tensor is

$$H_{\sigma\mu} = H^{\beta}{}_{\sigma\beta\mu} \quad (4.59a)$$

The curvature scalar is

$$\begin{aligned} R' = g^{\sigma\mu}R'_{\sigma\mu} &= + \partial^\sigma\Gamma^{\beta}{}_{\sigma\beta} - \partial_\beta\Gamma^{\beta}{}_{\sigma}{}^{\sigma} + \Gamma^{\gamma}{}_{\beta\sigma}\Gamma^{\beta}{}_{\gamma}{}^{\sigma} - \Gamma^{\gamma}{}_{\mu\sigma}\Gamma^{\beta}{}_{\gamma\beta} + \partial^\sigma\Gamma^{2\beta}{}_{\sigma\beta} - \partial_\beta\Gamma^{2\beta}{}_{\sigma}{}^{\sigma} + \Gamma^{2\gamma}{}_{\beta\sigma}\Gamma^{2\beta}{}_{\gamma}{}^{\sigma} - \\ &\quad - \Gamma^{2\gamma\sigma}{}_{\sigma}\Gamma^{2\beta}{}_{\gamma\beta} + \Gamma^{\gamma}{}_{\beta\sigma}\Gamma^{2\beta}{}_{\gamma}{}^{\sigma} - \Gamma^{\gamma\sigma}{}_{\sigma}\Gamma^{2\beta}{}_{\gamma\beta} + \Gamma^{2\gamma}{}_{\beta\sigma}\Gamma^{\beta}{}_{\gamma}{}^{\sigma} - \Gamma^{2\gamma\sigma}{}_{\sigma}\Gamma^{\beta}{}_{\gamma\beta} \\ &= g^{\sigma\mu}(R^{1\beta}{}_{\sigma\beta\mu} + R^{2\beta}{}_{\sigma\beta\mu}) \end{aligned} \quad (4.60)$$

An additional curvature scalar is

$$H = g^{\sigma\mu}H_{\sigma\mu} \tag{4.60a}$$

4.6 Separation of the Lagrangian Gravitational and SU(3) Sectors

In this section we will separate the gravitational and strong interaction parts of the total lagrangian which we form as an extension of the usual Einstein lagrangian. The major aspect of the extension is the introduction of higher derivative terms in a manner that can be handled by canonical lagrangian methods. We also separate the Ricci tensor into two parts in order to use pseudoQuantization field theory to implement canonical lagrangian methods and to introduce the flat space metric $\eta^{\sigma\mu}$ by a Higgs Mechanism. The constant flat space metric is usually an assumed quantity. But its close relation to the quantum field $g^{\sigma\mu}$ suggests that it could be generated by the same Higgs Mechanism that generates particle mass constants.[40]

Therefore we assume the lagrangian density:

$$\mathcal{L} = g[aR'^{1}{}_{\sigma\mu}R'^{2\sigma\mu} + bR' + cg^{\sigma\mu}g^{2}{}_{\sigma\mu} + eg^{2\sigma\mu}g^{2}{}_{\sigma\mu} - dA^{2}{}_{\mu}A^{2\mu}] \tag{4.61}$$

Since we wish to apply it cosmologically, and within hadrons, where the spinor connections are normally negligible we set $B^{1}{}_{\nu\mu} = B^{2}{}_{\nu\mu} = 0$ and find

$$\begin{aligned}\mathcal{L} &= \sqrt{g}[a(R'^{1}{}_{SU(3)\sigma\mu} + R'^{1}{}_{G\sigma\mu})(R'^{2}{}_{SU(3)}{}^{\sigma\mu} + R'^{2}{}_{G}{}^{\sigma\mu}) + bR' + c\,g^{\sigma\mu}g^{2}{}_{\sigma\mu} - dA^{2}{}_{\mu}A^{2\mu}]\\ &= \sqrt{g}[a(R'^{1}{}_{SU(3)\sigma\mu}R'^{2}{}_{SU(3)}{}^{\sigma\mu} + R'^{1}{}_{G\sigma\mu}R'^{2}{}_{G}{}^{\sigma\mu} + R'^{1}{}_{G\sigma\mu}R'^{2}{}_{SU(3)}{}^{\sigma\mu} + R'^{1}{}_{SU(3)\sigma\mu}R'^{2}{}_{G}{}^{\sigma\mu}) +\\ &\quad + bg^{\sigma\mu}(R'^{1}{}_{G}{}^{\beta}{}_{\sigma\beta\mu} + R'^{2}{}_{G}{}^{\beta}{}_{\sigma\beta\mu}) + cg^{\sigma\mu}g^{2}{}_{\sigma\mu} + eg^{2\sigma\mu}g^{2}{}_{\sigma\mu} - dA^{2}{}_{\mu}A^{2\mu}]\end{aligned} \tag{4.62}$$

Since there are no strong interaction fields in 'empty' space and gravity is neglgible within hadrons, we can drop the interaction terms between these interactions obtaining:

$$\begin{aligned}\mathcal{L} &= \sqrt{g}[a(R'^{1}{}_{SU(3)\sigma\mu}R'^{2}{}_{SU(3)}{}^{\sigma\mu} + R'^{1}{}_{G\sigma\mu}R'^{2}{}_{G}{}^{\sigma\mu}) + bg^{\sigma\mu}(R'^{1}{}_{G}{}^{\beta}{}_{\sigma\beta\mu} + R'^{2}{}_{G}{}^{\beta}{}_{\sigma\beta\mu}) + cg^{\sigma\mu}g^{2}{}_{\sigma\mu} + eg^{2\sigma\mu}g^{2}{}_{\sigma\mu} -\\ &\quad - dA^{2}{}_{\mu}A^{2\mu}]\\ &= \mathcal{L}_{SU(3)} + \mathcal{L}_{G}\end{aligned} \tag{4.63}$$

where

$$\mathcal{L}_{SU(3)} = \sqrt{g}[aR'^{1}{}_{SU(3)\sigma\mu}R'^{2}{}_{SU(3)}{}^{\sigma\mu} - dA^{2}{}_{\mu}A^{2\mu}] \tag{4.64}$$

$$\mathcal{L}_{G} = \sqrt{g}[aR'^{1}{}_{G\sigma\mu}R'^{2}{}_{G}{}^{\sigma\mu} + bg^{\sigma\mu}(R'^{1}{}_{G}{}^{\beta}{}_{\sigma\beta\mu} + R'^{2}{}_{G}{}^{\beta}{}_{\sigma\beta\mu}) + cg^{\sigma\mu}g^{2}{}_{\sigma\mu} + eg^{2\sigma\mu}g^{2}{}_{\sigma\mu}] \tag{4.65}$$

$$\mathcal{L}_{G} = \sqrt{g}[aR'^{1}{}_{G\sigma\mu}R'^{2}{}_{G}{}^{\sigma\mu} + bH + cg^{\sigma\mu}g^{2}{}_{\sigma\mu} + eg^{2\sigma\mu}g^{2}{}_{\sigma\mu}] \tag{4.65a}$$

Thus $\mathcal{L}_{SU(3)}$ is the dominant interaction within hadrons, and $\mathcal{L}_{G}$ is the dominant interaction in space within the framework of this discussion.

One may ask why we relate these two interactions. We gave a general reason earlier; they both share the unique feature of being exact symmetries. Another reason is that they both

---

[40] Blaha (2016c).

have coupling constants with dimension: G and the coefficient of the Strong Interaction linear potantial term.

More reasons that will emerge in the following chapters are: 1) a possible relationship between the constants appearing in eq. 4.65 and in eq. 4.69 below; and 2) a possible relationship between parts of these interactions. (Questions have been raised about the gravitational potential and the Strong Interaction effective potential.)

We now introduce a strong interaction coupling constant f with

$$A^1_{\ \mu} \rightarrow fA^{1\mu} \qquad (4.66)$$
$$A^2_{\ \mu} \rightarrow fA^{2\mu}$$

leading to

$$F^1_{\ \kappa\mu} = \partial A^1_{\ \mu}/\partial x^\kappa - \partial A^1_{\ \kappa}/\partial x^\mu + if[A^1_{\ \kappa}, A^1_{\ \mu}] \qquad (4.67)$$
$$F^2_{\ \kappa\mu} = \partial A^2_{\ \mu}/\partial x^\kappa - \partial A^2_{\ \kappa}/\partial x^\mu + if[A^2_{\ \kappa}, A^2_{\ \mu}] + if[A^1_{\ \kappa}, A^2_{\ \mu}] + if[A^2_{\ \kappa}, A^1_{\ \mu}]$$

Then the Strong Interaction lagrangian density terms are:[41]

$$\mathcal{L}_{SU(3)} = \sqrt{g}[afR'^1_{\ SU(3)\sigma\mu}R'^2_{\ SU(3)}{}^{\sigma\mu} - df^2A^2_{\ \mu}A^{2\mu}] \qquad (4.68)$$

We approximate g = 1 within hadrons. Thus eq. 4.68 becomes

$$\mathcal{L}_{SU(3)} = afR'^1_{\ SU(3)\sigma\mu}R'^2_{\ SU(3)}{}^{\sigma\mu} - df^2A^2_{\ \mu}A^{2\mu} \qquad (4.69)$$
$$= \zeta R'^1_{\ SU(3)\sigma\mu}R'^2_{\ SU(3)}{}^{\sigma\mu} - \varsigma A^2_{\ \mu}A^{2\mu} \qquad (4.70)$$

where we set

$$\zeta = af = \tfrac{1}{2} \qquad (4.71)$$
$$\varsigma = df^2 = \tfrac{1}{2}\lambda^2 \qquad (4.72)$$

We note that

$$b = (16\pi G)^{-1} \qquad (4.73)$$

in accord with the usual Einstein dynamic equations.

---

[41] The form is virtually identical to S. Blaha, Phys. Rev. D**11**, 2921 (1974) and Blaha's 1976 Gravity Research Foundation Essay (Honorable Mention). See Appendices A and D.

5. SU(3) Strong Interaction Dynamics

5.1 16-Dimensional Complex-Valued Coordinates

Earlier in section 4.4 we showed that the 32-dimensional real-value space corresponding to the vierbein $l^{\mu ai}(x)$ described in section 4.2 can be mapped to a 16-dimensional complex-valued space. This space can contain our 4-dimensional universe with real-valued coordinates due to the well-known map of our universe into 10-dimensional space described in Eddingon (1952). Since we choose, in earlier books, to take our universe to have complex-valued coordinates the 'Eddington' map requires a 10-dimensional complex-valued space. Thus our 16-dimensional complex universe, that we have called the Megaverse, easily contains our universe, and probably other universes, as surfaces within it.

Since our universe has complex-valued coordinates[42] we require complex-valued arguments for the Strong Interaction gauge fields that are consistent with the form of the coordinates of quarks with which they interact:

$$\begin{array}{l}\text{Real-valued Energies} \\ \text{Complex-valued spatial Momenta: } \mathbf{x} = \mathbf{x}_r + i\mathbf{x}_i\end{array} \tag{5.1}$$

This form of the coordinates is implemented after the extraction of the dynamical equations from the lagrangian.

On the other hand, due to the factorization of the lagrangian into separate gravitational and Strong Interaction parts we can use complex-valued coordinates in the gravitational sector without regard to the restrictions of eq. 5.1.

5.2 Strong Interaction Lagragian Terms

The Strong Interaction lagrangian (eq. 4.70) is

$$\mathcal{L}_{SU(3)} = ½F^{1}{}_{\kappa\mu}F^{2\kappa\mu} - ½\lambda^2 A^{2}{}_{\mu}A^{2\mu} \tag{5.2}$$

It is equal to eq. 17 of S. Blaha, Phys. Rev. D**11**, 2921 (1974) (appendix A) except for an additional term $[A^2{}_\kappa, A^2{}_\mu]$ that does not affect the conclusions of the paper.

Since the paper essentially contains a complete description of our Strong Interaction theory we refer the reader to the paper in Appendix A and a predecessor paper in Appendix B.

---

[42] In earlier work we showed that the four known species of fermions: charged leptons, neutral leptons, up-type quarks and down-type quarks, require complex space-time coordinates and the complex Lorentz group. In particular, quarks of both species require real-valued energies and complex-valued spatial momenta of the form $\mathbf{x} = \mathbf{x}_r + i\mathbf{x}_i$. We call particles of this type *complexons*.

There are a few 'modest' changes required to bring the paper into complete agreement with our current theory:

1. Eq. 30 of the paper must be modified to

$$F^2{}_{\kappa\mu} = \partial A^2{}_\mu/\partial x^\kappa - \partial A^2{}_\kappa/\partial x^\mu + \mathbf{if[A^2{}_\kappa, A^2{}_\mu]} + if[A^1{}_\kappa, A^2{}_\mu] + if[A^2{}_\kappa, A^1{}_\mu] \qquad (5.3)$$

with the addition of the term $if[A^2{}_\kappa, A^2{}_\mu]$. There is also a trivial change of notation of coupling constant from 'g' to 'f'.

2. Eqs. 6 and 18 should have the interaction term expanded to

$$g\not{A}^1 \rightarrow g(\not{A}^1 + \not{A}^2) \qquad (5.4)$$

and similarly in eq. 20 in Appendix B. Eqs. 38 – 41 of Appendix A directly show that the additional interaction term leads to a gluon propagator[43] $<A^1 + A^2, A^1 + A^2> = 2<A^1, A^2> + <A^1, A^1>$, and introduces a 1/r term in the potential part of the gluon propagator.

As a result the effective gluon propagator combines eqs. 38 and 39 (Appendix A) leading to the gluon propagator between quarks:

$$g_{\mu\nu}\delta_{ab}P[\lambda^2/k^4 - 1/k^2] \qquad (5.5)$$

up to a constant factor.[44]

This, in turn, *explicitly* leads to a Strong Interaction potential of the form

$$V(r) = -2f^2/r + f^2\lambda^2 r \qquad (5.6)$$

Naturally one can expect perturbative corrections to eq. 5.6 in higher order in f. However, as will be seen in the next subsection, the apparent relative smallness of f suggests eq. 5.6 is a good approximation to the inter-quark interaction.

5.3 Coupling Constant Values

The "Cornell group" developed an apparently satisfactory[45] charmed quark bound state spectrum in 1974-5 using a combination of a linear and a 1/r potential as the strong interaction. In a recent fit[46] they gave the potential energy:

---

[43] Eqs. 40-41 in Appendix A.

[44] This propagator is taken in Principal value to avoid potential unitarity problems. This topic is described in detail in Appendices A and B.

[45] As did a Harvard group.

[46] E. J. Eichten, K. Lane, and C. Quigg, arXiv:hep-ph/ 0206018 (2002). See this paper for references to earlier work by the "Cornell group" and the "Harvard group" as well as papers by other researchers.

$$V(r) = -\kappa/r + r/a_{Cornell}^2 \tag{5.7}$$

where $\kappa = 0.61$, $a_{Cornell} = 2.38\ \text{GeV}^{-1}$ and the charmed quark mass was 1.84 GeV. Substituting these values in eq. 5.6 gives

$$f = \sqrt{(\kappa/2)} = 0.552 \tag{5.8}$$
$$\lambda = 0.761\ \text{GeV} \tag{5.9}$$
$$d = \tfrac{1}{2}\lambda^2/f^2 = 0.950\ \text{GeV}^2 \tag{5.10}$$
$$a = 1/(2f) = 0.906 \tag{5.11}$$

The Cornell charmonium potential emerges directly from our theory. We note the closeness of a[47] and d to unity suggests that they may be one in value.

$$d = \tfrac{1}{2}\lambda^2/f^2 = 1.0 \quad ??? \tag{5.10'}$$
$$a = 1/(2f) = 1.0 \quad ??? \tag{5.11'}$$

and thus

$$f = \tfrac{1}{2} \qquad ???$$

Then the Cornell values would become

$$\kappa = \tfrac{1}{2}$$
$$a_{Cornell} = (f\lambda)^{-1} = 2.828\ \text{GeV}^{-1} \qquad ???$$

which are not very dissimular to the stated values, and might be accomodated by a shift in the mass of the charmed quark. We leave that to a future study.

We note that $f^2/4\pi = 0.024$ is only a factor of 3.3 more than the fine structure constant – approximately 1/137 = 0.0073. *Therefore perturbative corrections to our inter-quark potential may not be significant and our theory may be the correct theory of the strong interaction. The strong interaction potential in this charmonium fit suggests that the unperturbed potential of the theory presented in Appendices A and B may be a good approximation to the exact potential determined in perturbation theory.*

The smallness of the Strong Interaction terms in the Cornell group potential (eq. 5.7) looks puzzling at first glance. Why is it not large ("Strong")? We believe the strength of the Strong Interaction does not originate in the value of the coupling constant but rather in the linear potential term which provides confinement. The Cornell group potential's 'small' coupling constant then becomes understandable within the context of Strong Interaction

5.4 Implied Constraint on Gravitational Constant

Based on the above Cornell fit data we find that gravitational constants appearing in eqs. 4.63 and 4.65 are determined. Eq. 4.71 implies

$$a = 1/(2f) = 0.906 \tag{5.12}$$

[47] A value of a = 1 would make the coefficient of the 'quartic' term $R'^{1}{}_{G\sigma\mu}R'^{2}{}_{G}{}^{\sigma\mu}$ in eq. 4.65 unity.

We note

$$b = (16\pi G)^{-1} \tag{4.73}$$

from standard Gravity theory. Thus eq. 4.65 is determined except for the 'cosmological constant' c:

$$\mathcal{L}_G = \surd g[aR'^1{}_{G\sigma\mu}R'^2{}_G{}^{\sigma\mu} + bg^{\sigma\mu}(R'^1{}_G{}^\beta{}_{\sigma\beta\mu} + R'^2{}_G{}^\beta{}_{\sigma\beta\mu}) + cg^{\sigma\mu}g^2{}_{\sigma\mu} + eg^{2\sigma\mu}g^2{}_{\sigma\mu}] \tag{4.65}$$

We will see if eq. 4.65 leads to a satisfactory gravitational potential, modified from the Newtonian 1/r potential, in chapter 6.

6. Gravity Sector – A Slightly Modified Gravity at Ultra-Small Distances

In the previous chapter we determined the dynamics of the Strong Interaction sector of this unified Gravity-Strong Interaction theory that constitutes a subsector of our Theory of Everything. In this chapter we will develop the gravitation sector of this unified subsector.[48] We will see that the theory has higher derivative dynamic equations but unlike the Strong Interaction sector, which yields color (quark-gluon) confinement, the gravitation dynamic equations do not have confinement – but do yield a modified form of gravity at ultra-short distances of the order of the Planck length. The modification of gravity implied by our theory is consistent with the need for Dark Matter described in our Theory of Everything (Blaha (2015a) and (2016c)).[49] A MOND theory[50] would require a major change in General Relativity and is not part of the theory described here.

6.1 Gravitation Sector

Our gravity sector has two metric fields, $g_{\mu\nu}$ and $g^2{}_{\mu\nu}$ derived from the unified formalism described in chapter 4. Some of the relevant gravitation equations found in chapter 4 are:

$$H^\sigma{}_{\nu\mu} = \Gamma^\sigma{}_{\nu\mu} + \Gamma^{2\sigma}{}_{\nu\mu} \tag{4.43a}$$

$$H^\beta{}_{\sigma\nu\mu} = \partial_\mu H^\beta{}_{\sigma\nu} - \partial_\nu H^\beta{}_{\sigma\mu} + H^\gamma{}_{\nu\sigma} H^\beta{}_{\gamma\mu} - H^\gamma{}_{\mu\sigma} H^\beta{}_{\gamma\nu} \tag{4.50a}$$

$$H_{\sigma\mu} = H^\beta{}_{\sigma\beta\mu} \tag{4.59a}$$

$$H = g^{\sigma\mu} H_{\sigma\mu} \tag{4.60a}$$

---

[48] There are other higher derivative theories – some with two metrics, and some with a metric plus vector plus scalar field formulation. The present work is based on a unification of Strong and Gravity sectors and a totally different formalism. Some significant references are: M. Milgrom, Phys. Rev. **D80**, 123536 (2009); C. Skordis et al, Phys. Rev. Lett. **96**, 011301 (2006); R. H. Sanders, Astrophysical Journal **480**, 492 (1997); and references therein; J. D. Bekenstein, Phys. Rev. **D70**, 083509 (2004) and references therein; J-P. Bruneton, Phys. Rev. **D76**, 124012 (2007) with a higher derivative gravity and metric, vector, scalar fields. See also references within these articles.

[49] I. Ferreras et al, Phys. Rev. Lett., **96**, 011301 (2006) shows the need for both Dark Matter and MOND based on studies of astrophysical data.

[50] Our gravity theory has aspects similar to the MOND theories described in A. Balakin et al, Phys. Rev. **D70**, 064027 (2004); H-S Zhao et al, Phys. Rev. **D82**, 103001 (2010); and references therein. However the conclusions are vastly different.

$$\mathcal{H} = R'^1 + R'^2$$

$$\mathcal{L}_G = \sqrt{g}[aR'^1{}_{G\sigma\mu}R'^2{}_G{}^{\sigma\mu} + bH + cg^{\sigma\mu}g^2{}_{\sigma\mu} + eg^{2\sigma\mu}g^2{}_{\sigma\mu}] \quad (4.65a)$$

where a, b, c, and d are constants with

$$a = 1/(2f) = 0.906 \quad (5.11)$$

and

$$b = (16\pi G)^{-1} \quad (4.73)$$

G is Newton's gravitational constant.

The lagrangian dynamic equations following from eq. 4.65a are difficult. We consequently will examine the weak gravitation limiting case where we can approximate the two metrics with

$$g_{\mu\nu} \cong \eta_{\mu\nu} + h_{\mu\nu} \quad (6.1)$$
$$g^2{}_{\mu\nu} \cong \eta_{\mu\nu} + h^2{}_{\mu\nu} \quad (6.2)$$

where

$$|h_{\mu\nu}| << 1 \quad (6.3)$$
$$|h^2{}_{\mu\nu}| << 1$$

Using the relations

$$\partial_\mu h^\mu{}_\nu = \tfrac{1}{2}\partial_\nu h^\mu{}_\mu \quad (6.4)$$
$$\partial_\mu h^{2\mu}{}_\nu = \tfrac{1}{2}\partial_\nu h^{2\mu}{}_\mu \quad (6.5)$$

and neglecting higher order terms in $h_{\mu\nu}$ and $h^2{}_{\mu\nu}$ we find

$$R'^1{}_{\mu\nu} = \tfrac{1}{2}[\Box h_{\mu\nu} - \partial_\mu\partial_\lambda h^\lambda{}_\nu - \partial_\nu\partial_\lambda h^\lambda{}_\mu + \partial_\nu\partial_\mu h^\lambda{}_\lambda]$$
$$\cong \tfrac{1}{2}\Box h_{\mu\nu} \quad (6.6)$$
$$R'^1 \cong \tfrac{1}{2}\Box h_\mu{}^\mu$$

$$R'^2{}_{\mu\nu} = \tfrac{1}{2}[\Box h^2{}_{\mu\nu} - \partial_\mu\partial_\lambda h^{2\lambda}{}_\nu - \partial_\nu\partial_\lambda h^{2\lambda}{}_\mu + \partial_\nu\partial_\mu h^{2\lambda}{}_\lambda]$$
$$\cong \tfrac{1}{2}\Box h^2{}_{\mu\nu} \quad (6.7)$$
$$R'^2 \cong \tfrac{1}{2}\Box h^2{}_\mu{}^\mu$$

Substituting in eq. 4.65a above we find the *effective quadratic* part of the lagrangian (in $h_{\mu\nu}$ and $h^2{}_{\mu\nu}$) is

$$\mathcal{L}_G = \sqrt{g}[a\Box h_{\sigma\mu}\Box h^{2\sigma\mu}/4 + \tfrac{1}{2}b(\partial_\alpha h^{\sigma\mu}\partial^\alpha h_{\sigma\mu} + \partial_\alpha h^{2\sigma\mu}\partial^\alpha h^2{}_{\sigma\mu}) + c(4 + \eta^{\sigma\mu}h^2{}_{\sigma\mu} + h^{\sigma\mu}\eta_{\sigma\mu} + h^{\sigma\mu}h^2{}_{\sigma\mu}) +$$
$$+ e(2\eta^{\sigma\mu}h^2{}_{\sigma\mu} + h^{2\sigma\mu}h^2{}_{\sigma\mu}) + \tfrac{1}{2}(h_{\mu\nu} + h^2{}_{\mu\nu})T^{\mu\nu}] \quad (6.8)$$

Using partial integrations, we find the standard technique for determining the equations of motion from a lagrangian for independent variations with respect to $h_{\mu\nu}$ and $h^2{}_{\mu\nu}$ yields

$$a\Box^2 h^{2\mu\nu}/4 + \tfrac{1}{2}b\Box h^{\mu\nu} + c(\eta^{\mu\nu} + h^{2\mu\nu}) + \tfrac{1}{2}T^{\mu\nu} = 0 \qquad (6.9)$$

$$a\Box^2 h^{\mu\nu}/4 + \tfrac{1}{2}b\Box h^{2\mu\nu} + c(\eta^{\mu\nu} + h^{\mu\nu}) + 2e(\eta^{\mu\nu} + h^{2\mu\nu}) + \tfrac{1}{2}T^{\mu\nu} = 0 \qquad (6.10)$$

The term $c\eta^{\mu\nu}$ can be viewed as part of the total energy-momentum tensor $T'^{\mu\nu}$:

$$T'^{\mu\nu} = T^{\mu\nu} + 2c\eta^{\mu\nu} \qquad (6.10a)$$

It plays a role similar to the Cosmological Constant. Subtracting the equations we find

$$a\Box^2 h^{2\mu\nu}/4 - a\Box^2 h^{\mu\nu}/4 + \tfrac{1}{2}b\Box h^{\mu\nu} - \tfrac{1}{2}b\Box h^{2\mu\nu} + c(h^{2\mu\nu} - h^{\mu\nu}) - 2e(\eta^{\mu\nu} + h^{2\mu\nu}) = 0 \qquad (6.11)$$

and thus

$$[a\Box^2 - \tfrac{1}{2}b\Box + (4c - 8e)]h^{2\mu\nu} = 8e\eta^{\mu\nu} + a\Box^2 h^{\mu\nu} - \tfrac{1}{2}b\Box h^{\mu\nu} + 4ch^{\mu\nu} \qquad (6.12)$$

Therefore we determine the metric equation for $h^{2\mu\nu}$

$$h^{2\mu\nu} = [a\Box^2 - \tfrac{1}{2}b\Box + (4c - 8e)]^{-1}[8e\eta^{\mu\nu} + a\Box^2 h^{\mu\nu} - \tfrac{1}{2}b\Box h^{\mu\nu} + 4ch^{\mu\nu}] \qquad (6.13)$$

Substituting in eq. 6.9 we obtain

$$[a\Box^2/4 + c][a\Box^2 - \tfrac{1}{2}b\Box + (4c - 8e)]^{-1}[8e\eta^{\mu\nu} + a\Box^2 h^{\mu\nu} - \tfrac{1}{2}b\Box h^{\mu\nu} + 4ch^{\mu\nu}] + \tfrac{1}{2}b\Box h^{\mu\nu} + \tfrac{1}{2} T'^{\mu\nu} = 0 \qquad (6.14)$$

We now define an energy-momentum tensor

$$T''^{\mu\nu} = [16ec/(4c - 8e)] + T'^{\mu\nu} \qquad (6.15)$$

with the result eq. 6.14 becomes

$$\{[a\Box^2/4 + c][a\Box^2 - \tfrac{1}{2}b\Box + (4c - 8e)]^{-1}[a\Box^2 - \tfrac{1}{2}b\Box + 4c] + \tfrac{1}{2}b\Box\}h^{\mu\nu} = -\tfrac{1}{2}T''^{\mu\nu} \qquad (6.16)$$

6.2 A Post-Newtonian Gravity Potential

Eq. 6.16 leads to the form of a new gravitational potential:

$$V(\mathbf{x}) = -\int d^3k\, V(\mathbf{k}) \exp(i\mathbf{k}\cdot\mathbf{x})/(2\pi)^3 \qquad (6.17)$$

where

$$V(\mathbf{k}) = \{[a\mathbf{k}^4/4 + c][a\mathbf{k}^4 - \tfrac{1}{2}b\mathbf{k}^2 + (4c - 8e)]^{-1}[a\mathbf{k}^4 - \tfrac{1}{2}b\mathbf{k}^2 + 4c] + \tfrac{1}{2}b\mathbf{k}^2\}^{-1} \qquad (6.18)$$

$$= 4[a\mathbf{k}^4 - \tfrac{1}{2}b\mathbf{k}^2 + (4c - 8e)]/\{[a\mathbf{k}^4 + 2b\mathbf{k}^2 + 4c][a\mathbf{k}^4 - \tfrac{1}{2}b\mathbf{k}^2 + 4c] - 16eb\mathbf{k}^2\}$$

$$= 4[a\mathbf{k}^4 - ½b\mathbf{k}^2 + (4c - 8e)]/\{[a^2\mathbf{k}^8 + 3ab\mathbf{k}^6/2 + (8ac - b^2)\mathbf{k}^4 + (6bc - 16eb)\mathbf{k}^2 + 16c^2\}$$

The denominator can be factored into a product of four factors

$$(\mathbf{k}^2 - m_1^2)(\mathbf{k}^2 - m_2^2)(\mathbf{k}^2 - m_3^2)(\mathbf{k}^2 - m_4^2) \quad (6.19)$$

by solving the quartic (biquadratic) equation in $\mathbf{k}^2$

$$[a^2\mathbf{k}^8 + 3ab\mathbf{k}^6/2 + (8ac - b^2)\mathbf{k}^4 + (6bc - 16eb)\mathbf{k}^2 + 16c^2 = 0 \quad (6.20)$$

The exact solution may be found using the method found in Chrystal (1961) pp 550-553. Here we shall present an approximate, physically interesting, solution here based on the largeness of $b = (16\pi G)^{-1}$ and leave the exact solution to a future work should a need develop for it.

The form of the resulting exact form of $V(\mathbf{k})$ is

$$V(\mathbf{k}) = 4[a\mathbf{k}^4 + ½b\mathbf{k}^2 + (4c - 8e)]/\{a^2 (\mathbf{k}^2 - m_1^2)(\mathbf{k}^2 - m_2^2)(\mathbf{k}^2 - m_3^2)(\mathbf{k}^2 - m_4^2)\} \quad (6.21)$$

Substituting in eq. 6.17 we obtain a sum of gradiants of Yukawa potentials:

$$V(\mathbf{r}) =[a\nabla^4 - ½b\nabla^2 + (4c - 8e)] \sum_{i=1}^{4} \alpha_i\, e^{-m_i r}/(4\pi r) \quad (6.22)$$

where α_i are constants.

The observant reader may inquire: Where is the well-known 1/r Newtonian potential? We suggest that possibly the graviton has a *very* small mass of the order of the Planck mass or smaller, which is thereby undetectable –perhaps even at the cosmological level. The validity of this answer may be seen if we recognize the largeness of the constant b, and approximate eq. 6.18 accordingly to obtain the Newtonian gravitational potential:

$$V(\mathbf{k}) \cong -2b\mathbf{k}^2/\{-b^2\mathbf{k}^4\} = 2/(b\mathbf{k}^2) = 32\pi G/\mathbf{k}^2 \rightarrow V(\mathbf{x}) = -G/|\mathbf{x}| \quad (6.23)$$

6.2.1 Potential Due to the Secondary Metric $h^{2\mu\nu}$

The lagrangian eq. 4.65 indicates that the secondary metric $h^{2\mu\nu}$ also couples to the energy momentum tensor. Thus there is a secondary potential

$$V^2(\mathbf{x}) = -\int d^3k\, V^2(\mathbf{k}) \exp(i\mathbf{k}\cdot\mathbf{x})/(2\pi)^3 \quad (6.24)$$

that exerts a gravitational force. From eq. 6.13 we see

$$[a\nabla^4 + ½b\nabla^2 + (4c - 8e)]V^2(\mathbf{x}) = [a\nabla^4 + ½b\nabla^2 + 4c]V(\mathbf{x}) \quad (6.25)$$

Applying eq. 6.25 to eq. 6.24 we find

$$
\begin{aligned}
V^2(\mathbf{k}) &= [a\mathbf{k}^4 - \tfrac{1}{2}b\mathbf{k}^2 + (4c - 8e)]^{-1}[a\mathbf{k}^4 - \tfrac{1}{2}b\mathbf{k}^2 + 4c]V(\mathbf{k}) \\
&= 4[a\mathbf{k}^4 - \tfrac{1}{2}b\mathbf{k}^2 + 4c]/\{[a\mathbf{k}^4 + 2b\mathbf{k}^2 + 4c][a\mathbf{k}^4 - \tfrac{1}{2}b\mathbf{k}^2 + 4c] - 16eb\mathbf{k}^2\} \\
&= 4[a\mathbf{k}^4 - \tfrac{1}{2}b\mathbf{k}^2 + 4c]/\{[a^2\mathbf{k}^8 + 3ab\mathbf{k}^6/2 + (8ac - b^2)\mathbf{k}^4 + (6bc - 16eb)\mathbf{k}^2 + 16c^2\}
\end{aligned}
\tag{6.26}
$$

with $V(\mathbf{k})$ determined by eq. 6.18.

If the constant b dominates then

$$V^2(\mathbf{k}) \cong -2b\mathbf{k}^2/\{[-b^2\mathbf{k}^4\} = 2\mathbf{k}^2/b = 32\pi G/\mathbf{k}^2 \rightarrow V^2(\mathbf{x}) = -G/|\mathbf{x}| \tag{6.27}$$

Taking account of the factor of ½ eminating from the term $\tfrac{1}{2}(h_{\mu\nu} + h^2{}_{\mu\nu})T^{\mu\nu}$ in eq. 6.8 we find the sum

$$\tfrac{1}{2}(V(\mathbf{x}) + V^2(\mathbf{x})) = -G/|\mathbf{x}| \tag{6.28}$$

yields the Newtonian potential.

6.2.2 Graviton Feynman Propagators

The Feynman propagators for the 'free' field case considered above have pole terms with negative signs that lead to unitarity problems. Thus the propagators must be taken in Principal value to avoid potential unitarity problems just as we indicated in the previous chapter for Strong Interaction gluons. This topic is described in detail in Appendices A and B. The theory then becomes an Action-at-a-distance theory analogous to Feynman and Wheeler's Action-at-a-Distance Electrodynamics.

The lagrangian equation eq. 4.65 implies the non-zero equal time commutators:

$$[h_{\mu\nu}(x), h_{\lambda\sigma}(y)]$$

and

$$[h^2{}_{\mu\nu}(x), h_{\lambda\sigma}(y)]$$

and the corresponding non-zero Feynman propagators

$$G_{\mu\nu\lambda\sigma} = \langle 0|T(h_{\mu\nu}(x), h_{\lambda\sigma}(y))|0\rangle$$

and

$$G^2{}_{\mu\nu\lambda\sigma} = \langle 0|T(h^2{}_{\mu\nu}(x), h_{\lambda\sigma}(y))|0\rangle$$

The Feynman propagators, by eq. 6.21 and 6.26, have the form

$$i\,G_{\mu\nu\lambda\sigma}(z) = -iP \int d^4k\, e^{-ik\cdot z}\, b_{\mu\nu\lambda\sigma}(k)4[ak^4 - \tfrac{1}{2}bk^2 + (4c - 8e)]/\{a^2(k^2 - m_1{}^2)(k^2 - m_2{}^2)(k^2 - m_3{}^2)(k^2 - m_4{}^2)\}/(2\pi)^4$$

$$i\,G^2{}_{\mu\nu\lambda\sigma}(z) = -iP \int d^4k\, e^{-ik\cdot z}\, b^2{}_{\mu\nu\lambda\sigma}(k)4[ak^4 + \tfrac{1}{2}bk^2 + 4c]/\{[a^2k^8 + 3abk^6/2 + (8ac - b^2)k^4 + (6bc - 16eb)k^2 + 16c^2\}/(2\pi)^4$$

where P indicates principal value,[51] and $b_{\mu\nu\lambda\sigma}(k)$ and $b^2_{\mu\nu\lambda\sigma}(k)$ are polynomials in k. Appendices A and B have discussions of the need for principal value propagators to avoid unitarity problems.

6.3 Solution for the Case of a Massless Graviton

If c = 0 in the eq. 4.65 lagrangian then

$$\mathcal{L}_G = \sqrt{g}[aR'^1{}_{G\sigma\mu}R'^2{}_G{}^{\sigma\mu} + bH + eg^{2\sigma\mu}g^2{}_{\sigma\mu}] \qquad (4.65a')$$

and the quartic equation eq. 6.20 factors to a cubic form that yields masses as solutions of the equation:

$$a^2\mathbf{k}^2[\mathbf{k}^6 + 3ab\mathbf{k}^4/2 - b^2\mathbf{k}^2 - 16be] = 0 \qquad (6.29)$$

It leads to a Newtonian potential –G/r plus three massive 'graviton' terms that cannot conspire to approximate the MOND potential. The cubic equation has the form

$$x^3 + Bx^2 + Cx + D = 0 \qquad (6.30)$$

with $x = \mathbf{k}^2$

$$B = 3ab/2 \qquad C = -b^2 \qquad D = -16be$$

Setting x = y + w, and letting w = –B/3 we can transform eq. 6.30 to

$$y^3 + C'y + D' = 0 \qquad (6.31)$$

where

$$C' = C - B^2/3 \qquad D' = D + 2B^2/27 - BC/3$$

This cubic equation has three roots m_1^2, m_2^2 and m_3^2. Since we require the three roots to be real value on physical grounds, we can use Cardan's Irreducible Case[52] to find the real roots expressed in trigonometric functions:

$$\begin{aligned} m_1^2 &= -2\rho^{1/3}\cos(\theta/3) \\ m_2^2 &= -2\rho^{1/3}\cos((2\pi + \theta)/3) \\ m_3^2 &= -2\rho^{1/3}\cos((4\pi + \theta)/3) \end{aligned} \qquad (6.32)$$

where

$$\rho = (-C')^{3/2}/27^{1/2}$$
$$\cos\theta = 27^{1/2}D'/[2(-C')^{3/2}]$$

or

$$\rho = [b^2 + 3(ab/2)^2]^{3/2}/27^{1/2} \qquad (6.33)$$
$$\cos\theta = 27^{1/2}(-16be + (ab)^2/6 + ab^3)/2)/[2(b^2 + 3(ab/2)^2)^{3/2}]$$

---

[51] The use of a principal value is required due to the denominator. It is thus similar to

[52] See Chrystal (1961) p. 549.

Thus eq. 6.21 becomes

$$V(\mathbf{k}) = 4[a\mathbf{k}^4 + ½b\mathbf{k}^2 - 8e]/\{a^2\mathbf{k}^2(\mathbf{k}^2 - m_1^2)(\mathbf{k}^2 - m_2^2)(\mathbf{k}^2 - m_3^2)\} \quad (6.34)$$

If we assume b is the dominant large parameter then

$$\rho \sim b^3[1 + 3a^2/4]^{3/2}/27^{½} \sim b^3[2.04/5.2] = 0.39b^3 \quad (6.35)$$
$$\cos\theta \sim 27^{½}(a/4)/(1 + 3a^2/4)^{3/2} \sim 0.574$$

assuming a ~ 0.906. Then all masses are of the order of the Planck mass ~ √b

$$m_1^2 \sim -1.39b \quad (6.36)$$
$$m_2^2 = 0.71b$$
$$m_3^2 = 1.16b$$

Note that one mass m_1 is tachyonic. Since our Theory of Everything[53] supports tachyons and since there is some evidence for tachyonic neutrinos we do not necessarily regard this graviton mass as a negative.

6.3.1 Eliminating the Possible Tachyonic Graviton

An alternative is to set the constant e to cause the tachyonic term to cancel. The numerator $a\mathbf{k}^4 + ½b\mathbf{k}^2 - 8e$ of eq. 6.34 has roots determined by

$$ax^2 + (b/2)x - 8e = 0 \quad (6.37)$$

The roots are

$$\mathbf{k}^2 = [-b/2 \pm (b^2/4 + 32ae)^{½}]/2a \quad (6.38)$$

To cancel the tachyonic root we must have

$$-1.39b = [-b/2 - (b^2/4 + 32ae)^{½}]/2a$$

This determines the constant e to be

$$e = b^2[(2.78a - 1/2)^2 - 1/4]/(32a) = 0.13b^2 \quad (6.39)$$

As a result eq. 6.34 becomes

$$V(\mathbf{k}) = (2/a)\{\mathbf{k}^2 - [-b/2 + (b^2/4 + 32ae)^{½}]/(2a)\}/\{\mathbf{k}^2(\mathbf{k}^2 - m_2^2)(\mathbf{k}^2 - m_3^2)\} \quad (6.40)$$

---

[53] See Blaha (2015a) for example for evidence of tachyon neutrinos.

For large distances ($\mathbf{k}^2 \to 0$) we have the potential

$$V(\mathbf{k}) \to (2/a)(1.39b)/\{\mathbf{k}^2 m_2{}^2 m_3{}^2\} = (2/a)(1.39b)/\{\mathbf{k}^2(0.71b)(1.16b)\} \qquad (6.40)$$
$$= 3.72/(b\mathbf{k}^2) \approx 2(32\pi G)/\mathbf{k}^2 \to -2G/|\mathbf{x}|$$

with the difference being due to roundoff error.

The secondary metric leads to a the potential

$$V^2(\mathbf{k}) = [a\mathbf{k}^4 - \tfrac{1}{2}b\mathbf{k}^2 \ -8e]^{-1}[a\mathbf{k}^4 - \tfrac{1}{2}b\mathbf{k}^2]V(\mathbf{k}) \qquad (6.41)$$

$$V^2(\mathbf{k}) \to \tfrac{1}{2}b\mathbf{k}^2 V(\mathbf{k})/(8e) = \mathbf{k}^2/[16(.13b)]\cdot 3.72/(b\mathbf{k}^2) \to V^2(\mathbf{x}) = 1.79/b^2 \simeq 0$$

and the total gravity potential for large distance is the Newtonian potential:

$$V_{total}(\mathbf{k}) \simeq \tfrac{1}{2}(V(\mathbf{k}) + 0) = (32\pi G)/\mathbf{k}^2 \to -G/|\mathbf{x}| \qquad (6.42)$$

The large masses m_2 and m_3 of the additional graviton particles means that they only affect gravity at distances below the Planck distance. These additional poles generate very small distance Yukawa potentials. Thus the effective gravitational potential for the present case has the form:

$$V(\mathbf{r}) \sim -G/r + a_1 G e^{-m_2 r}/r + a_2 G e^{-m_3 r}/r \qquad (6.43)$$

where a_1 and a_2 are constants.

A change in the value of the constant a ***cannot*** *change the values of m_2 and m_3 in such a way as to make them have small masses which could then affect the large distance behavior of gravity and lead to a MOND-like gravity potential.*

7. Reduction of Complex General Relativity to Real-valued General Relativity Using a U(4) Reality Group and the Gravitational Higgs Mechanism

7.1 Complex General Relativity

This chapter[54] reduces complex-valued General Relativity to real-valued General Relativity through the introduction of a U(4) Reality group.[55] Its starting point is the complex General Relativity sector of the unified theory of the Strong Interactions and General Relativity prsented in chapter 4.

An additional motivation for complex General Relativity is provided in Blaha (2015a). In that work, and earlier books, we showed complex General Relativity like real General Relativity has its origin in 4-dimensional space-time. In chapters 2 and 4 of Blaha (2015a) we showed that 4-dimensional space-time had its 'ultimate' origin in Asynchronous Logic. We then showed that flat space-time was complex. Generalizing to curved space-time it is necessary for consistency to assume a *complex* curved space-time that is made real using the Reality group. Applying Ockham's Razor and Leibniz's Minimax Principle we find it the simplest choice yielding compatibility between complex flat space-time and its curved space-time extension.

We can then derive the Einstein dynamic equations of General Relativity for both real-valued and complex-valued coordinates.

7.2 Reality Group Transformations in Curved Complex Space-Time

The General Relativity U(4) Reality group enables any 4-dimensional complex vector to be transformed to a real-valued vector or *vice versa.* If we define local Reality group transformations then we can transform the set of coordinates of a reference frame to real values. In this section we specify the matrix form of the generators of U(4) and relate them to the U(4) subgroups. Then we will define tetrad (vierbein) forms of the generators in an inertial reference frame for later use in this chapter in defining the dynamic equations of the new Complex General Relativity.

It will become apparent that the use of the Reality group to map Complex General Relativity to real General Relativity will reveal a correspondence with the Extended Standard Model that we previously used to establish a Theory of Everything.

The Extended Standard Model sector can then be viewed as being defined in an inertial reference frame at each point in a curved space-time. If we imagine the curved space-time as

---

[54] This chapter is an extract from Blaha (2015b) with some changes.

[55] This Reality group differs from the U(4) Generation group and the U(4) Layer group that we introduced in earlier books.

smoothly beconing flat then the curved space-time definition of the Extended Standard Model then extends throughout flat space-time.

7.2.1 4-Dimensional Representation of the Subgroup Generators of the U(4) Reality Group

Four dimensional representations of some of the sixteen generators of U(4) were defined in Blaha (2011c). These generators can be put in the form of the non-commuting subgroups' 16 generators for SU(3), SU(2), U(1), SU(2) and U(1). Since we wish to use these generators to implement transfomations in 4-dimensional space-time we must define 4×4 matrix *reducible* representations of each subgroup's set of generators. Thus the sixteen 4×4 generator τ_k (for k = 1, … , 16) representations are:

SU(3) – a $\underline{3}\oplus\underline{1}$ SU(3) representation of 8 generators
SU(2) – a $\underline{3}\oplus\underline{1}$ SU(2) representation of 3 generators
U(1) – a $\underline{1}\oplus\underline{1}\oplus\underline{1}\oplus\underline{1}$ U(1) representation of 1 generator
SU(2) – another $\underline{3}\oplus\underline{1}$ SU(2) representation of 3 generators
U(1) – another $\underline{1}\oplus\underline{1}\oplus\underline{1}\oplus\underline{1}$ U(1) representation of 1 generator

7.2.2 Scalar Fields of the Reality group

Since we develop a unified theory of everything we can define scalar fields using the notation of the Extended Standard Model in Blaha (2015a):

SU(3) – Φ_i for i = 1, 2, …, 8 (7.1)
SU(2) – Φ_i for i = 1, 2, 3
U(1) – Φ_0
SU(2) – Φ'_i for i = 1, 2, 3
U(1) – Φ'_0

It is convenient to relabel the above fields as $\Phi_k(x)$ for k = 1, 2, …, 16. It will also be convenient to label the 4×4 generator matrices as listed above as τ_k in the same respective order.

The combined treatment of the various Reality group symmetries in a 4×4 matrix representation may seem at first strange. However their combination within one representation (based on the four dimensionality of space-time) for use in defining a part of the complex general relativistic transformations of Complex General Relativity is acceptable and, more importantly, provides the needed complexity of the theory. We note that the transformation defined below in eq. 7.2 below is a matrix but the symmetries of the various fields $\Phi_k(x')$ is subsumed in the inner product with U(4) generator matrices τ_k and so U(x), defined below, becomes simply a 4×4 U(4) matrix dependent on x.

The U(4) generators may be used to define a U(4) general coordinate transformation in four dimensions which corresponds to the Reality group:

$$U(x) = \exp[i \sum_k g_k\Phi_k(x')\tau_k] \equiv e^{i\theta(x)} \quad (7.2)$$

where the sum is for k = 1, …, 16. There are 16 U(4) coupling constants, $\{g_k\}$.

7.2.3 Tetrad (Vierbein) Form of the 4×4 Reality Group Transformations

In this subsection we will map 4×4 Reality group transformations (eq. 7.2) to the form of General Relativistic transformations using *tetrads* (*vierbeins*).[56] Then we will construct Complex General Relativistic transformations, the complex affine connection, and then dynamical gravitational equations.

The *vierbein* formalism begins with the Equivalence Principle that allows us to define an inertial coordinate system in the neighborhood of any point Z in space-time. We will use the notation $\varsigma^\alpha(Z)$ to denote the inertial coordinates at Z. We define a tetrad or vierbein as

$$v^\alpha{}_\mu(x) = (\partial\varsigma^\alpha(x)/\partial x^\mu)_{x=Z} \quad (7.3)$$

In a neighborhood of Z we can invert the relation between ς and x to define an inverse

$$w^\mu{}_\alpha(x) = (\partial x^\mu(\varsigma)/\partial\varsigma^\alpha)_{x=X} \quad (7.4)$$

such that

$$w^\mu{}_\alpha(x)v^\alpha{}_\nu(x) = \delta^\mu{}_\nu \quad (7.5)$$
$$w^\mu{}_\beta(x)v^\alpha{}_\mu(x) = \delta^\alpha{}_\beta \quad (7.6)$$

In real-valued General Relativity all *tetrads* are real-valued. In Complex General Relativity a *tetrad* $v^\alpha{}_\mu(x)$ is complex-valued.

The metric at a curved space-time point X is defined in terms of *tetrads* as

$$g_{\rho\sigma}(x) = \eta_{\alpha\beta}\, v^\alpha{}_\rho(x)v^\beta{}_\sigma(x) \quad (7.7)$$
$$g^{\rho\sigma}(x) = \eta^{\alpha\beta}\, w^\rho{}_\alpha(x)w^\sigma{}_\beta(x) \quad (7.8)$$

The inverse of a *tetrad* transformation can also be expressed as

$$w_\beta{}^\nu(x) = v_\beta{}^\nu(x) = \eta_{\beta\alpha}g^{\nu\mu}(x)v^\alpha{}_\mu(x) \quad (7.9)$$

Then a *tetrad* and its inverse satisfy the relations

$$v^\alpha{}_\mu(x)v_\beta{}^\mu(x) = \delta^\alpha{}_\beta \quad (7.10)$$

[56] In this chapter it seemed more reasonable to use a slightly different set of vierbein symbols for better readability.

and

$$v^{\alpha}{}_{\mu}(x)v_{\alpha}{}^{\nu}(x) = \delta^{\nu}{}_{\mu} \tag{7.11}$$

7.2.3.1 Transformations of Tetrads

There are two general types of space-time transformations that can be performed on a tetrad.

1. A complex-valued (possibly real-valued) General Relativistic coordinate transformation:

$$v'^{\alpha}{}_{\mu}(x) = \partial x^{\nu}/\partial x'^{\mu}\, v^{\alpha}{}_{\nu}(x) \tag{7.12}$$

2. A complex-valued, local *Lorentzian transformation*

$$v'^{\beta}{}_{\mu}(x) = \Lambda(x)^{\beta}{}_{\alpha}\, v^{\alpha}{}_{\mu}(x) \tag{7.13}$$

where $\Lambda(x)^{\beta}{}_{\alpha}$ is an element of a subset of the local Complex Lorentz Group to be specified later.

7.2.3.2 Local Lorentzian Formalism for Tetrads

The local Lorentzian transformations $\Lambda(x)^{\beta}{}_{\alpha}$ consist of local Lorentz transformations that are real-valued, and complex-valued Lorentz transformations. Both types of transformations satisfy the orthogonality condition:

$$\eta_{\alpha\beta}\Lambda^{\alpha}{}_{\rho}(x)\Lambda^{\beta}{}_{\sigma}(x) = \eta_{\rho\sigma} \tag{7.14}$$

Thus the *tetrad* partakes of both local (position dependent) General Relativistic transformations and local Lorentzian transformations.

7.2.3.3 Tetrad (Vierbein) Form of the 4×4 Reality Group Transformations

In eq. 7.2 we defined Reality group matrix transformations. Using the matrix form of this definition, and tetrads, we see that we can express a Reality transformation in an inertial coordinate system in the neighborhood of a point Z in space-time. Thus

$$\begin{aligned} U^{\mu}{}_{\nu}(x) &\equiv w^{\mu}{}_{a}(x)U^{a}{}_{b}(x)v^{b}{}_{\nu}(x) = w^{\mu}{}_{a}(x)[e^{i\theta(x)}]^{a}{}_{b}v^{b}{}_{\nu}(x) \\ &= w^{\mu}{}_{a}(x)[\textstyle\sum_{n} (i\theta(x))^{n}/n!]^{a}{}_{b}v^{b}{}_{\nu}(x) \\ &= w^{\mu}{}_{a}(x)v^{a}{}_{\nu}(x) + w^{\mu}{}_{a}(x)i\theta(x)^{a}{}_{b}v^{b}{}_{\nu}(x) + \ldots \\ &= \delta^{\mu}{}_{\nu} + [\theta(x)]^{\mu}{}_{\nu} + [\theta(x)]^{\mu}{}_{\alpha}[\theta(x)]^{\alpha}{}_{\nu}/2 + \ldots \end{aligned} \tag{7.15}$$

where we have transformed the terms within the expansion of the exponentiated matrix into local Lorentz tensors, and where

$$[\theta(x)]^{\mu}{}_{\alpha} = w^{\mu}{}_{a}(x)\,[\theta(x)]^{a}{}_{b}\,v^{b}{}_{\alpha}(x) \tag{7.15a}$$

with a and b being 4×4 matrix column and row indices of $[\tau_k]^a{}_b$ (See eq. 7.2.). The local Lorentzian tensorial form of U(x) can be used to develop Complex General Coordinate transformations as we do in the next section.

For later use we note that the inverse of eq. 7.15 is

$$U^{-1\alpha}{}_{\mu}(x) = w^{\alpha}{}_{a}(x)U^{-1a}{}_{b}(x)v^{b}{}_{\mu}(x) \tag{7.16}$$

with

$$U^{-1\alpha}{}_{\mu}(x)U^{\mu}{}_{\nu}(x) = \delta^{\alpha}{}_{\nu} \tag{7.17}$$

where

$$U^{-1a}{}_{b}(x) = [U^{\dagger}(x)]^{a}{}_{b} \tag{7.18}$$

with † signifying hermitean conjugate.

7.3 Structure of Complex General Coordinate Transformations

Complex General Coordinate transformations can be uniquely factored into products of two terms. They have the form

$$\partial x''^{\nu}(x)/\partial x^{\mu} = U(x'')^{\nu}{}_{\beta}\,\partial x'^{\beta}(x)/\partial x^{\mu} \tag{7.19}$$

where

$$x''^{\nu}(x) = U(x'')^{\nu}{}_{\beta}x'^{\beta}$$
$$x'^{\mu}(x) = U^{-1\mu}{}_{b}(x'')\,x''^{b}$$

where $U(x')^{\nu}{}_{\beta}$ is complex and where $\partial x'^{\beta}(x)/\partial x^{\mu}$ is a purely real General Coordinate transformation.

We define

$$U(x'')^{\mu}{}_{\nu} = w^{\mu}{}_{a}(x'')\big[\exp\big(i\textstyle\sum_k g_k\Phi_k(x'')\tau_k\big)\big]^{a}{}_{b}\,v^{b}{}_{\nu}(x'') \tag{7.20}$$

$$U^{-1}(x'')^{\mu}{}_{\nu} = w^{\mu}{}_{a}(x'')\big[\exp\big(-i\textstyle\sum_k g_k\Phi_k(x'')\tau_k\big)\big]^{a}{}_{b}\,v^{b}{}_{\nu}(x'') \tag{7.21}$$

where the constants g_k are real, and Φ_k and τ_k are hermitean. The uniqueness of the factorization follows from the Reality group (and the U(4)) property that any complex 4-vector can be uniquely mapped to any specified real 4-vector.

Given the factorization (eq. 7.19) it becomes possible to separate the affine connection correspondingly. Then the dynamical gravitational equations of Complex General Relativity can be made to exhibit their Φ_k field dependence in a manner analogous to Higgs fields in the Extended Standard Model. And a part of a Theory of Everything becomes evident.

7.4 Complex Affine Connection

The structure of a complex general coordinate transformation (eq. 7.19) enables us to calculate its affine connection for later use in determining the covariant derivative, and the dynamic equations. First the transformation to the real-valued x' coordinates from inertial coordinates is

$$\Gamma^{\sigma}{}_{\lambda\mu}\,(x') = \partial x'^{\sigma}/\partial\varsigma^{\rho}\;\partial^2\varsigma^{\rho}/\partial x'^{\lambda}\partial x'^{\mu} \tag{7.22}$$

Next the Reality group transformation has the affine connection

$$\Gamma^{\sigma}{}_{\lambda\mu}\,(x'') = \partial x''^{\sigma}/\partial\varsigma^{\rho}\;\partial^2\varsigma^{\rho}/\partial x''^{\lambda}\partial x''^{\mu} \tag{7.23}$$

which becomes

$$\Gamma^{\sigma}{}_{\lambda\mu}\,(x'') = \partial x''^{\sigma}/\partial x'^{\beta}\;\partial x'^{\beta}(\varsigma)/\partial\varsigma^{\rho}\;\partial/\partial x''^{\mu}[\partial\varsigma^{\rho}/\partial x'^{\alpha}\;\partial x'^{\alpha}/\partial x''^{\lambda}] \tag{7.24}$$

Using eq. 7.22 we find eq. 7.24 has the form

$$\Gamma^{\sigma}{}_{\lambda\mu}(x'') = \partial x''^{\sigma}/\partial x'^{\beta}\;\partial x'^{\alpha}/\partial x''^{\lambda}\;\partial x'^{\gamma}/\partial x''^{\mu}\;\Gamma^{\beta}{}_{\alpha\gamma}(x') + \partial x''^{\sigma}/\partial x'^{\beta}\;\partial^2 x'^{\beta}/\partial x''^{\lambda}\partial x''^{\mu} \tag{7.25}$$

Next substituting the Reality group transformation

$$x''^{\nu}(x) = U(x'')^{\nu}{}_{\beta}x'^{\beta}$$
$$x'^{\mu}(x) = U^{-1}(x'')^{\mu}{}_{\beta}\,x''^{\beta} \tag{7.26}$$

together with

$$\partial x''^{\sigma}/\partial x'^{\beta} = \partial[U(x'')^{\sigma}{}_{\alpha}x'^{\alpha}]/\partial x'^{\beta} = U(x'')^{\sigma}{}_{\beta} + x'^{\alpha}\,\partial U(x'')^{\sigma}{}_{\alpha}/\partial x'^{\beta} \tag{7.27}$$
$$\partial x'^{\sigma}/\partial x''^{\beta} = \partial[U^{-1}(x'')^{\sigma}{}_{\alpha}x''^{\alpha}]/\partial x''^{\beta} = U^{-1}(x'')^{\sigma}{}_{\beta} + x''^{\alpha}\,\partial U^{-1}(x'')^{\sigma}{}_{\alpha}/\partial x''^{\beta} \tag{7.28}$$

we find the second term in eq. 7.25 is the Reality fields affine connection

$$\Gamma_V{}^{\sigma}{}_{\lambda\mu}(x'') = \partial[U(x'')^{\sigma}{}_{\alpha}x'^{\alpha}]/\partial x'^{\beta}\;\partial\{\partial[U^{-1}(x'')^{\beta}{}_{\alpha}x''^{\alpha}]/\partial x''^{\lambda}\}/\partial x''^{\mu} \tag{7.29}$$

7.5 Complex Curvature Tensor and Complex Einstein Equation of the New Complex General Relativity

The complex space-time Riemann-Christoffel curvature tensor is

$$R^{\rho}{}_{\mu\nu\sigma}(x'') \equiv \partial\Gamma^{\rho}{}_{\mu\nu}(x'')/\partial x''^{\sigma} - \partial\Gamma^{\rho}{}_{\mu\sigma}(x'')/\partial x''^{\nu} + \Gamma^{a}{}_{\mu\nu}(x'')\Gamma^{\rho}{}_{\sigma a}(x'') - \Gamma^{a}{}_{\mu\sigma}(x'')\Gamma^{\rho}{}_{\nu a}(x'') \tag{7.30}$$

as can be shown by following the standard steps in its derivation. (We note that the algebra of the tensor manipulations is the same as in the usual derivation of General Relativistic quantities and equations.)

The Complex General Relativistic Einstein Equation is

$$R_{\mu\nu}(x'') - \tfrac{1}{2}\, g_{\mu\nu}\, R(x'') = -8\pi G\, T_{\mu\nu} \tag{7.31}$$

where G is Newton's gravitational constant ($6.674 \times 10^{-11}\ m^3kg^{-1}s^{-2}$), $T_{\mu\nu}$ is the energy-momentum tensor, the Ricci tensor is

$$R_{\mu\nu}(x'') = R^{a}{}_{\mu a\nu}(x'') \tag{7.32}$$

and the curvature scalar is

$$R(x'') = g^{\mu\nu}R_{\mu\nu}(x'') \tag{7.33}$$

with $g^{\mu\nu} = g^{\mu\nu}(x'')$.

7.6 Approximate Form of the New Complex General Relativity Dynamic Equations

The introduction of scalar fields, to embody the transformation of real-valued coordinates and expressions to complex values, creates a new form of Complex General Relativity if the scalar fields have dynamic equations, which together with the complex Einstein equation, are jointly solved. A solution of the complete set of equations is not possible presently.

Since the form of $\Gamma^{\sigma}{}_{\lambda\mu}(x'')$ as given in eqs. 7.25 – 7.29 is sufficiently complicated to make a solution of the new Einstein equation eq. 7.31 impossible currently, we will explore an approximete solution. It is possible to derive a leading order approximation if the scalar fields are sufficiently weak (with the result that the complex gravitational field is almost real-valued), and also if the gravitational field is also sufficiently weak. Then we can approximate $U(x'')^{\nu}{}_{\beta}$ and $U^{-1}(x'')^{\nu}{}_{\beta}$ of eqs. 7.20-7.21 with

$$U(x'')^{\nu}{}_{\beta} \approx \delta^{\nu}{}_{\beta} + i\sum_k g_k\Phi_k(x'')[\tau_k]^{\nu}{}_{\beta} \tag{7.34}$$

$$U^{-1}(x'')^{\nu}{}_{\beta} \approx \delta^{\nu}{}_{\beta} - i\sum_k g_k\Phi_k(x'')[\tau_k]^{\nu}{}_{\beta} \tag{7.35}$$

using

$$w^{\mu}{}_{a}(x'') \approx \delta^{\mu}{}_{a}$$

$$v^{b}{}_{\nu}(x'') \approx \delta^{b}{}_{\nu}$$

Then eq. 7.25 becomes

$$\Gamma^{\sigma}{}_{\lambda\mu}(x'') \approx \Gamma_{GR}{}^{\sigma}{}_{\lambda\mu}(x') + \Gamma_{V}{}^{\sigma}{}_{\lambda\mu}(x'') \tag{7.36}$$

to leading order in $V_{\mu k}(x'')$ where $\Gamma_{GR}{}^{\sigma}{}_{\lambda\mu}(x')$ is the usual general relativistic affine connection and where

$$\Gamma_{V}{}^{\sigma}{}_{\lambda\mu}(x'') \equiv U(x'')^{\sigma}{}_{\beta}\, \partial[U^{-1}(x'')^{\beta}{}_{\lambda}]/\partial x''^{\mu} \tag{7.37}$$

$$\approx -\tfrac{1}{2}\, i\{\textstyle\sum_k g_k\partial\Phi_k(x'')/\partial x''^{\mu}\, [\tau_k]^{\sigma}{}_{\lambda} + \sum_k g_k\partial\Phi_k(x'')\, /\partial x''^{\lambda}[\tau_k]^{\sigma}{}_{\mu}\}$$

in view of the forms of the Ricci tensor and the scalar curvature. The other terms lead to zero when used in the Einstein dynamic equation. In a more compact notation, we have

$$\Gamma_V{}^{\sigma}{}_{\lambda\mu}(x'') \approx iA^{\sigma}{}_{\lambda\mu}$$

with $\lambda\mu$ symmetry where

$$A^{\sigma}{}_{\lambda\mu} = -\tfrac{1}{2}\{\textstyle\sum_k g_k\partial\Phi_k(x'')/\partial x''^{\mu}\, [\tau_k]^{\sigma}{}_{\lambda} + \sum_k g_k\partial\Phi_k(x'')\, /\partial x''^{\lambda}[\tau_k]^{\sigma}{}_{\mu}\} \quad (7.37a)$$
$$= -\tfrac{1}{2}\{\partial/\partial x''^{\mu}\, \varphi^{\sigma}{}_{\lambda} + \partial\, /\partial x''^{\lambda}\, \varphi^{\sigma}{}_{\mu}\} \quad (7.37b)$$

with

$$\varphi^{\sigma}{}_{\mu} = \textstyle\sum_k g_k\Phi_k(x'')[\tau_k]^{\sigma}{}_{\mu} \quad (7.37c)$$

7. 7 Approximate Particle Motion in a Weak Gravity Field and in Weak Reality Fields Φ_k

The motion of a particle in a freely falling coordinate system ς^{ρ} satisfies

$$d^2\varsigma^{\rho}/d\tau^2 = 0$$

It can be expressed in an arbirary coordinate system x^{ρ} in the form

$$d^2x^{\rho}/d\tau^2 + \Gamma^{\rho}{}_{\mu\nu}\, dx^{\mu}/d\tau\, dx^{\nu}/d\tau = 0$$

For a slowly moving particle in a weak stationary gravitational, and in weak scalar, fields we can approximate the particle dynamic equation with

$$d^2x^{\rho}/d\tau^2 + \Gamma^{\rho}{}_{00}\, (dt/d\tau)^2 = 0$$

neglecting spatial derivative terms. Substituting eqs. 7.36 and 7.37 we obtain

$$d^2x^{\rho}/d\tau^2 + [\Gamma^{\rho}{}_{00}(x) + i\Gamma_V{}^{\rho}{}_{00}(x)]\, (dt/d\tau)^2 = 0$$

where we approximate x" = x' = x with

$$\Gamma_V{}^{\rho}{}_{00}(x) = A^{\rho}{}_{00}$$

if all the $V_{0k}(x)$ fields are slowly varying with time. Thus

$$d^2x^{\rho}/d\tau^2 + [\Gamma^{\rho}{}_{00}(x) + i\, A^{\rho}{}_{00}](dt/d\tau)^2 = 0$$

The equation for t is

$$d^2t/d\tau^2 + i\, A^{\rho}{}_{00}\, (dt/d\tau)^2 = 0$$

under the assumption that the gravitational field is weak. Its solution is

$$t - t_0 = -\ln(\tau - \tau_0)/\{i\, A^{\rho}{}_{00}\}$$

assuming the time dependence of $V_{0k}(x)$ is negligible. As a result

$$dt/d\tau = -\ 1/\{i(\tau - \tau_0)A^{\rho}{}_{00}\}$$

Assuming a weak field approximation so that

$$g_{\mu\nu} \approx \eta_{\mu\nu} + h_{\mu\nu}$$

we see the gravitational affine connection can be approximated with

$$\Gamma^{\rho}{}_{00}(x) = -\tfrac{1}{2}\, \eta^{\rho\nu}\, \partial h_{00}/\partial x^{\nu}$$

resulting in the spatial equation

$$d^2\mathbf{x}/d\tau^2 = \tfrac{1}{2}\, (dt/d\tau)^2\, \nabla\, h_{00}$$

which after simple manipulations becomes

$$i\, A^{\rho}{}_{00}\, d\mathbf{x}/dt + d^2\mathbf{x}/dt^2 = \tfrac{1}{2}\, \nabla\, h_{00}$$

In the limit where $\partial\Phi_k(x'')/\partial x''^{\mu} \approx 0$ for all k, since it is driven by gravitation (as we will see later) we find the form of the Newtonian gravitational potential emerges:

$$d^2\mathbf{x}/dt^2 = \tfrac{1}{2}\, \nabla\, h_{00}$$

corresponding to

$$d^2\mathbf{x}/dt^2 = \tfrac{1}{2}\, \nabla\, \varphi(\mathbf{x})$$

of Newton.

7.8 Approximate Riemann-Christoffel Curvature Tensor and Einstein Dynamic Equations

The complex space-time Riemann-Christoffel curvature tensor (eq. 7.30) then is approximately (by eqs. 7.35-37)

$$R^{\rho}{}_{\mu\nu\sigma}(x'') \approx \partial\Gamma^{\rho}{}_{\mu\nu}(x')/\partial x'^{\sigma} - \partial\Gamma^{\rho}{}_{\mu\sigma}(x')\,/\partial x'^{\nu} + \partial\Gamma_{V}{}^{\rho}{}_{\mu\nu}(x'')/\partial x''^{\sigma}\ - \partial\Gamma_{V}{}^{\rho}{}_{\mu\sigma}(x'')/\partial x''^{\nu} +$$

$$+ [\Gamma^{a}{}_{\mu\nu}(x') + \Gamma_V{}^{a}{}_{\mu\nu}(x'')]\,[\Gamma^{\rho}{}_{\sigma a}(x') + \Gamma_V{}^{\rho}{}_{\sigma a}(x'')] -$$
$$- [\Gamma^{a}{}_{\mu\sigma}(x') + \Gamma_V{}^{a}{}_{\mu\sigma}(x'')]\,[\Gamma^{\rho}{}_{\nu a}(x') + \Gamma_V{}^{\rho}{}_{\nu a}(x'')] \tag{7.38}$$

using

$$\partial/\partial x''^{\sigma} = \partial x'^{\beta}/\partial x''^{\sigma}\,\partial/\partial x'^{\beta} \approx [U^{-1}(x'')^{\beta}{}_{\sigma} + x''^{a}\,\partial U^{-1}(x'')^{\beta}{}_{\sigma}/\partial x''^{\beta}]\partial/\partial x'^{\beta} \approx \partial/\partial x'^{\sigma} \tag{7.39}$$

to leading order. Rearranging terms we find

$$R^{\rho}{}_{\mu\nu\sigma}(x'') \approx R_{GR}{}^{\rho}{}_{\mu\nu\sigma}(x') + R_V{}^{\rho}{}_{\mu\nu\sigma}(x'') + R_{GRV}{}^{\rho}{}_{\mu\nu\sigma}(x') \tag{7.40}$$

where

$$R_{GR}{}^{\rho}{}_{\mu\nu\sigma}(x') = \partial\Gamma^{\rho}{}_{\mu\nu}(x')/\partial x'^{\sigma} - \partial\Gamma^{\rho}{}_{\mu\sigma}(x')\,/\partial x''^{\nu} + \Gamma^{a}{}_{\mu\nu}(x')\Gamma^{\rho}{}_{\sigma a}(x') - \Gamma^{a}{}_{\mu\sigma}(x')\Gamma^{\rho}{}_{\nu a}(x') \tag{7.41}$$

$$R_V{}^{\rho}{}_{\mu\nu\sigma}(x'') = \partial\Gamma_V{}^{\rho}{}_{\mu\nu}(x'')/\partial x''^{\sigma} - \partial\Gamma_V{}^{\rho}{}_{\mu\sigma}(x'')\,/\partial x''^{\nu} + \Gamma_V{}^{a}{}_{\mu\nu}(x'')\Gamma_V{}^{\rho}{}_{\sigma a}(x'') -$$
$$- \Gamma_V{}^{a}{}_{\mu\sigma}(x'')\Gamma_V{}^{\rho}{}_{\nu a}(x'') \tag{7.42}$$

$$R_{GRV}{}^{\rho}{}_{\mu\nu\sigma}(x') = \Gamma^{a}{}_{\mu\nu}(x')\,\Gamma_V{}^{\rho}{}_{\sigma a}(x') + \Gamma_V{}^{a}{}_{\mu\nu}(x')\Gamma^{\rho}{}_{\sigma a}(x') - \Gamma^{a}{}_{\mu\sigma}(x')\Gamma_V{}^{\rho}{}_{\nu a}(x') -$$
$$- \Gamma_V{}^{a}{}_{\mu\sigma}(x')\Gamma^{\rho}{}_{\nu a}(x') \tag{7.43}$$

letting x'' ≈ x' in eq. 7.43. The Ricci tensor is

$$R_{\mu\nu} = R^{a}{}_{\mu a\nu} \approx R_{GR}{}^{a}{}_{\mu a\nu}(x') + R_V{}^{a}{}_{\mu a\nu}(x'') + R_{GRV}{}^{a}{}_{\mu a\nu}(x') \tag{7.44}$$

and the scalar curvature is

$$R(x'') = g^{\mu\nu}R_{\mu\nu}(x'') \approx g^{\mu\nu}[R_{GR}{}^{a}{}_{\mu a\nu}(x') + R_V{}^{a}{}_{\mu a\nu}(x'') + R_{GRV}{}^{a}{}_{\mu a\nu}(x')] \tag{7.45}$$

where

$$g^{\mu\nu} \approx g^{\mu\nu}(x') \approx g^{\mu\nu}(x'')$$

The Einstein dynamic equation for this new Complex General Relativity is superficially familiar:

$$R_{\mu\nu}(x'') - \tfrac{1}{2}\, g_{\mu\nu}\, R(x'') = -8\pi G\, T_{\mu\nu} \tag{7.46}$$

Eq. 7.46 (unlike the real General Relativistic equation) consists of two parts, a real part and an imaginary part, which together jointly determine the dynamics of real-valued gravitation and the dynamics of the scalar fields. The imaginary part consists of sixteen equations for the sixteen scalar fields coupled to gravitation. The real part in our approximation describes gravitation as the standard real-valued equation with no coupling to the scalar fields. The scalar fields' dynamic equations have terms coupling them to gravity.

Thus we have a theory combining gravitation with scalar fields that is similar to the Extended Standard Model where fermions are coupled to gauge fields (and at large energies appear to become a GUT with a U(4)⊗U(4) symmetry as described in chapter 3 of Blaha (2015b).)[57]

The real parts of the Ricci tensor and the scalar curvature *exactly*, in this approximation, yield the conventional *real-valued* Einstein dynamic equations

$$R_{GR\mu\nu}(x'') - \tfrac{1}{2}\, g_{\mu\nu}R_{GR}(x'') = -8\pi G\, T_{\mu\nu} \tag{7.47}$$

The approximate imaginary part of the Ricci tensor is

$$\begin{aligned} R_{I\mu\sigma} = \text{Im}\, R^{\alpha}{}_{\mu\alpha\sigma} \approx \text{Im}\, \{&\partial\Gamma_V{}^{\rho}{}_{\mu\rho}(x'')/\partial x''^{\sigma} - \partial\Gamma_V{}^{\rho}{}_{\mu\sigma}(x'')/\partial x''^{\rho} + \Gamma_V{}^{\alpha}{}_{\mu\rho}(x'')\Gamma_V{}^{\rho}{}_{\sigma\alpha}(x'') - \\ &- \Gamma_V{}^{\alpha}{}_{\mu\sigma}(x'')\Gamma_V{}^{\rho}{}_{\rho\alpha}(x'') + \Gamma_V{}^{\alpha}{}_{\mu\rho}(x'')\Gamma^{\rho}{}_{\sigma\alpha}(x') - \Gamma_V{}^{\alpha}{}_{\mu\sigma}(x'')\Gamma^{\rho}{}_{\rho\alpha}(x')\} \end{aligned} \tag{7.48}$$

The approximate imaginary part of the scalar curvature is

$$\begin{aligned} R_I = \text{Im}\, R = g^{\mu\sigma}R_{I\mu\sigma} = \text{Im}\{&\partial\Gamma_V{}^{\rho}{}_{\mu\rho}(x'')/\partial x''_{\mu} - \partial\Gamma_V{}^{\rho}{}_{\mu}{}^{\mu}(x'')/\partial x''^{\rho} + \Gamma_V{}^{\alpha}{}_{\mu\rho}(x'')\Gamma_V{}^{\rho\mu}{}_{\alpha}(x'') - \\ &- \Gamma_V{}^{\alpha}{}_{\mu}{}^{\mu}(x'')\Gamma_V{}^{\rho}{}_{\rho\alpha}(x'') + \Gamma_V{}^{\alpha}{}_{\mu\rho}(x'')\Gamma^{\rho\mu}{}_{\alpha}(x') - \Gamma_V{}^{\alpha}{}_{\mu}{}^{\mu}(x'')\Gamma^{\rho}{}_{\rho\alpha}(x')\} \end{aligned} \tag{7.49}$$

The imaginary part of the complex General Relativity Einstein equations

$$R_{I\mu\sigma} - \tfrac{1}{2} g_{\mu\nu}R_I(x'') = 0 \tag{7.50}$$

consists of a scalar particles' dynamic equation with interaction terms coupling to the real gravitational field with no direct coupling to the energy-momentum $T_{\mu\nu}$ (under the assumption $T_{\mu\nu}$ is real-valued although a tachyonic energy momentum density would imply a complex $T_{\mu\nu}$).

Subsituting eq. 7.37 for $\Gamma_V{}^{\rho}{}_{\mu\rho}(x'')$ we find

$$R_{I\mu\sigma} \approx i\partial A^{\rho}{}_{\mu\rho}/\partial x''^{\sigma} - i\partial A^{\rho}{}_{\mu\sigma}/\partial x''^{\rho} - A^{\alpha}{}_{\mu\rho}A^{\rho}{}_{\sigma\alpha} + A^{\alpha}{}_{\mu\sigma}A^{\rho}{}_{\rho\alpha} + iA^{\alpha}{}_{\mu\rho}\Gamma^{\rho}{}_{\sigma\alpha} - iA^{\alpha}{}_{\mu\sigma}\Gamma^{\rho}{}_{\rho\alpha}$$

and

$$R_I \approx i\partial A^{\rho}{}_{\mu\rho}/\partial x''_{\mu} - i\partial A^{\rho\mu}{}_{\mu}/\partial x''^{\rho} - A^{\alpha}{}_{\mu\rho}A^{\rho\mu}{}_{\alpha} + A^{\alpha}{}_{\mu}{}^{\mu}A^{\rho}{}_{\rho\alpha} + iA^{\alpha}{}_{\mu\rho}\Gamma^{\rho}{}_{\mu\alpha} - iA^{\alpha}{}_{\mu}{}^{\mu}\Gamma^{\rho}{}_{\rho\alpha} \tag{7.51}$$

---

[57] This has identical form to the derivation of Higgs fields from gauge fields described in Blaha (2015c) and (2016c).

Substituting eq. 7.37a we find the *linear terms* in the imaginary part of the Einstein equation (eq. 7.50) are[58]

$$\tfrac{1}{2}i[\partial^2\varphi^{\rho}{}_{\sigma}/\partial x''^{\rho}\partial x''^{\mu} - \partial^2\varphi^{\rho}{}_{\rho}/\partial x''^{\sigma}\partial x''^{\mu} + \tfrac{1}{2}g_{\mu\sigma}(\Box\varphi^{\rho}{}_{\rho} - \partial^2\varphi^{\rho}{}_{\beta}/\partial x''^{\rho}\partial x''_{\beta})] \qquad (7.52)$$

These terms plus the quadratic terms in eq. 7.50 constitute sixteen equations in the sixteen unknown fields $\Phi_k(x'')$. Since all terms in eq. 7.50 have only derivatives of $\Phi_k(x'')$, $\Phi_k(x'')$ is only specified up to an arbitrary constant c_k. We see

$$\Phi'_k(x'') = \Phi_k(x'') + c_k \qquad (7.53)$$

is also a solution of the imaginary Einstein equation for any choice of the 16 constants c_k.

7.9 The $\Phi_k(x'')$ Fields are the Higgs Fields

The freedom to shift the solutions $\Phi_k(x'')$ of the imaginary Einstein equation by a constant enables us to take these fields to be Higgs fields that can be used to give mass contributions to gauge bosons and fermions in Extended Standard Model symmetry breaking. Usually, as in Blaha (2015a), we simply insert kinetic terms and potential terms in the Extended Standard Model lagrangian. However, we propose, in a spirit of economy engendered by Ockham's Razor and Leibniz's MiniMax Principle, to take the imaginary Einstein lagrangian to be kinetic terms in The Theory of Everything lagrangian. We complete the Theory of Everything lagrangian by adding the Higgs potential energy terms to eq. 7.50 yielding the overall dynamic equations of the gravitational Higgs sector:

$$R_{I\mu\sigma} - \tfrac{1}{2}g_{\mu\nu}R_I(x'') = -8\pi i G_{Higgs}T_{Higgs\mu\sigma}(\Phi, V_{Higgs}(\Phi)) \qquad (7.54)$$

where G_{Higgs} is a constant, where the gravitational Higgs particle energy-momentum tensor $T_{Higgs\mu\sigma}$ is

$$T_{Higgs\mu\sigma}(\Phi, V_{Higgs}(\Phi)) = T_{HiggKinetics\mu\sigma}(\Phi) + \partial V_{Higgs}(\Phi)/\partial\Phi\ \partial\Phi/\partial x^{\mu}\ \partial\Phi(x)/\partial x^{\sigma} \qquad (7.54a)$$

and where $T_{HiggsKinetic\mu\sigma}(\Phi)$ is the kinetic part of the energy-momentum tensor for the gravitational Higgs fields.

---

58 Since the quadratic terms do not add to the significance of our discussion we defer their discussion, and the discussion of solutions, to a subsequent time.

The shift of gravitational Higgs fields to their vacuum expectation values is allowed due to the presence of only derivatives on the left side terms of eq. 7.54.

Consequently Gravitational Higgs particles can have a vacuum expectation value Φ_{k0} such that a gravitational Higgs field is a sum of a vacuum expectation value and a field $\varphi_k(x)$ dependent on x:

$$\Phi'_k(x) = \varphi_k(x) + \Phi_{k0} \tag{7.55}$$

7.10 Appearance of Gravitational Higgs Fields in Fermion Dynamic Equations

The fermion equations have Higgs terms that become masses (plus Higgs fields) after spontaneous breakdown. The introduction of a gravitational Higgs term in the fermion dynamic equations must be as a scalar term:

$$g^{\mu}{}_{\sigma}\varphi^{\sigma}{}_{\mu} = \sum_k g_k\Phi_k(x'')g^{\mu}{}_{\sigma}[\tau_k]^{\sigma}{}_{\mu} = \sum_k g_k\Phi_k(x'')[\tau_k]^{\sigma}{}_{\sigma}$$

The τ_k matrices are the generators of U(4). They have been transformed using the tetrad formalism (eq. 7.15a)

$$[\tau_k]^{\mu}{}_{\upsilon} = w^{\mu}{}_{a}(x)\,[\tau_k]^{a}{}_{b}\,v^{b}{}_{\upsilon}(x)$$

and so we find the trace of the generators

$$[\tau_k]^{\sigma}{}_{\sigma} = [\tau_k]^{a}{}_{a}$$

Since there are four diagonal generators in the 4-dimensional U(4) representation we find the gravitational Higgs term in each fermion dynamic equation is

$$g^{\mu}{}_{\sigma}\varphi^{\sigma}{}_{\mu} = \sum_k g_k\Phi_k(x'')$$

where the sum over k is over the k values of the four diagonal τ matrices. Inserting eq. 7.55 we find the gravitational Higgs term in each fermion dynamic equation is:

$$g^{\mu}{}_{\sigma}\varphi^{\sigma}{}_{\mu} = \sum_k g_k\,[\Phi_{k0} + \varphi_k(x)] \tag{7.55a}$$

yielding the mass term Φ_0

$$\sum_k g_k\Phi_{k0} = \Phi_{Grav0} \tag{7.55b}$$

which contains four gravitational Higgs particles vacuum expectation values.[59] Thus twelve of the gravitational Higgs particles need not have non-zero vacuum expectation values although they have dynamic significance due to the $\varphi_k(x)$ fields' dynamic equations.

7.11 Fermion Masses Contributions from Gravitational Higgs Particles

In chapter 2 of Blaha (2015b) and in Blaha (2015a) we discussed the fermion mass contributions from the vacuum expectation values of ElectroWeak and Dark ElectroWeak Higgs particles, and from vacuuu expectation values of Generation group Higgs particles (chapter 2 and chapter 16 of Blaha (2015a)). In this section we will complete the picture for fermion mass contributions by adding contributions from the Gravitational Higgs particles vacuuu expectation values that we described in this chapter.

Introducing Gravitational Higgs particles the lagrangian mass terms for the four normal and four Dark fermion species plus other contributions we find the Lagrangian terms are[60]

$$\begin{aligned}
\mathcal{L}^{Higgs}{}_{FermionMasses} = {} & \Sigma_{k,a,\alpha,\beta}\bar{\psi}_{kaL\alpha\delta}\eta_k m_{EW_{ka\alpha\beta}}\psi_{kaR\beta} + \Sigma_{k,a,\alpha,\beta}\bar{\psi}_{D_{kaL\alpha}}\eta_{Dk} m_{DEW_{ka\alpha\beta}}\psi_{D_{kaR\beta}} + && \text{ElectroWeak} \\
& + \Sigma_{k,a,\alpha,\beta}\bar{\psi}_{UkaL\alpha}\eta_{Uka} m_{Uka\alpha\beta}\psi_{UkaR\beta} + && \text{Generation} \\
& + \Sigma_{k,a,\alpha,\beta}\bar{\psi}_{DUkaL\alpha}\eta_{DUka} m_{DUka\alpha\beta}\psi_{DUkaR\beta} + && \text{Group U} \\
& + \Sigma_{k,g,\delta,\gamma}\bar{\psi}_{LkgL\delta}\eta_{Lg} m_{Lg\delta\gamma}\psi_{LkgR\gamma} + && \\
& + \Sigma_{k,g,\delta,\gamma}\bar{\psi}_{DLkgL\delta}\eta_{DLg} m_{DLg\delta\gamma}\psi_{DLkgR\gamma} + && \text{Layer Group L} \\
& + \Sigma_{k,a}\bar{\psi}_{GkaL}\eta_{Ga} m_{Gka}\psi_{GkaR} + \Sigma_{k,a}\bar{\psi}_{DGkaL}\eta_{DGa} m_{DGka}\psi_{DGkaR} + && \text{Gravitational} \\
& + \text{c.c.} && (5.56')
\end{aligned}$$

where the subscripts EW, D, U, L and G label ElectroWeak origin, D Dark type, U Generation group origin, L Layer group origin, and G Gravitational origin respectively. The fields labeled η (with subscripts) are Higgs fields that have non-zero vacuum expectation values.[61] The indices k label species – normal and Dark separately, g labels the (four) generations, and a labels the layers. The index δ and γ label *layer* rows and columns (with implicit sums over generations in the Layer group terms.) The Layer group mass contribution is the same for each fermion in each generation for each species in each layer. The matrices labeled m (with subscripts) are the complex constant mass matrices of species. The indices α, β = 1, … , 4 label *generation* rows and columns.

---

[59] The constants g_k can be set equal to a constant g_{grav}.

[60] Excerpted from Blaha (2016c).

[61] The Higgs fields η… in our pseudoquantum formulation are $\eta_{\ldots} = \varphi_{1\ldots}(x) + \varphi_{2\ldots}(x)$ as described earlier.

Eq. 5.56' contains the mass terms for the four layers of fermions in our Theory of Everything. *For each species and generation, the Layer group matrix terms mix the Layer mass contributions.* The three "upper" layers have terms with similar forms but with different mass values. These values are presumably very large. We expect that they are in the TeV and tens of TeVs ranges putting them probably out of range of the current CERN LHC.

Due to the weakness of the ultra-weak interaction, which is mitigated by the anticipated large vacuum expectation values, we expect significant mass cross terms in the mass matrixes of different layers.

The Generation group mass matrices cause mixing between the masses of the four generations in each species. The Layer group mass matrices cause mixing between masses of the four layers of each generation.

7.11.1 The Full Extended Standard Model Fermion Mass Matrices

Combining[62] the terms in eq. 5.56' for each species we obtain the total mass matrices *for each of the four layers* below. Each mass matrix can then be diagonalized to obtain the masses of the fermions within each of its species.[63]

Charged Lepton Species Total Mass Matrix

$$m_{etot} = m_{EWe} + m_{Ge} + m_{Le}$$

Neutral Lepton Species Mass Matrix

$$m_{\upsilon tot} = m_{EW\upsilon} + m_{G\upsilon} + m_{Lv}$$

Up-Type Quark Species Mass Matrix (for each color)

$$m_{utot} = m_{EWu} + m_{Uu} + m_{Gu} + m_{Lu}$$

Down-Type Quark Species Mass Matrix (for each color)

$$m_{dtot} = m_{EWd} + m_{Ud} + m_{Gd} + m_{Ld}$$

Dark Charged Lepton Species Total Mass Matrix

$$m_{Detot} = m_{DEWe} + m_{DGe} + m_{DLe}$$

Dark Neutral Lepton Species Mass Matrix

$$m_{D\upsilon tot} = m_{DEW\upsilon} + m_{DG\upsilon} + m_{DL\upsilon}$$

Dark Up-Type Quark Species Mass Matrix

$$m_{Dutot} = m_{DEWu} + m_{DUu} + m_{DGu} + m_{DLu}$$

Dark Down-Type Quark Species Mass Matrix

$$m_{Ddtot} = m_{DEWd} + m_{DUd} + m_{DGd} + m_{DLd}$$

We now note that the preceding formal development yields $m_{Ge} = m_{G\upsilon} = m_{Gu} = m_{Gd} = m_{DGe} = m_{DG\upsilon} = m_{DG\upsilon} = m_{DGu} = m_{DGd} = m_G$. The gravitational mass contribution to all fermions of all species is the same.

---

[62] Also excerpted from Blaha (2016c).

[63] Each Layer's mass contributions are shown.

Moreover, the gravitational contribution to each fermion mass sets the scale for all fermion masses (and secondarily of massive gauge bosons' masses) yielding the "principle" of Newton, Einstein and others that *inertial mass equals gravitational mass*.

NOTE: The generation group contributions, in the spontaneous breakdown that we described, appear only in quark and Dark quark mass matrices providing, possibly, a reason why quark masses are so much larger than lepton masses.

The mass matrices above can each be diagonalized in a manner similar to that of eqs. 16.50 and 16.51 in chapter 2 of Blaha (2015b) and in Blaha (2015a).

7.11.2 Inertial Mass Equals Gravitational Mass and a Local Arrow of Time

As we noted in previous books[64] the gravitational Higgs Mechanism contributions to all fermion masses implies the Principle: *Inertial Mass Equals Gravitational Mass* – a major independent principle until our work showed it followed from the gravitational Higgs Mechanism. It also introduces a *local* Arrow of Time, which is evident in particle interactions. Usually statistical arguments are used to establish a macroscopic arrow of time. The local Arrow of Time follows from our pseudoQuantum formalism for the Higgs Mehanism.

---

[64] For more detail see chapter 3 of Blaha (2016c).

Appendix A. Second Quantized Non-Abelian Theory of Quark Confinement

This refereed paper is S. Blaha, Phys. Rev. D**11**, 2921 (1974). Reprinted with the kind permission of Physical Review D.

Second-quantized non-Abelian field theory for hadrons with quark confinement and scaling deep-inelastic structure functions*

Stephen Blaha
Laboratory of Nuclear Studies, Cornell University, Ithaca, New York 14853
(Received 30 December 1974)

A four-dimensional second-quantized field theory with quarks bound by "colored" non-Abelian gluons is described which has the following properties: (1) the only physical particles are color singlets composed solely of quarks, (2) the deep-inelastic structure functions have Bjorken scaling, (3) gluon loops and Faddeev-Popov ghost loops are identically zero in any gauge, (4) Regge trajectories are apparently linear on a Chew-Frautschi plot, and (5) constituent motion within hadrons can be nonrelativistic.

I. INTRODUCTION

After a period of some skepticism the possibility that hadronic interactions might be understood within the framework of quantum field theory is again being seriously considered.[1] This is partly the result of the psychological climate created by the apparently successful unification of weak and electromagnetic interactions in a renormalizable field theory and partly the result of a greater appreciation of the variety of phenomena which can occur in field theories.

In this article we shall describe a field-theoretic model of hadron binding which has two major features: (1) Hadrons only occur as quark-antiquark or three-quark bound states, and (2) quarks behave as quasifree particles within hadrons. We assume that the suggestions of an internal symmetry called color[2] are correct and that the strong interaction consists of the exchange of colored Yang-Mills gluons. The nature of the interaction allows only color singlet states to occur in the gauge-invariant physical particle spectrum and consequently the first feature will be realized by choosing the color group to be SU(3). Since the (Schwinger) mechanism which produces this result is an infrared phenomenon, the second feature is not precluded and the model is essentially free in the ultraviolet region of the quark sector.

Our model is a non-Abelian version of a recently investigated Abelian field theory which had quark confinement and scaling electroproduction structure functions.[3] In that theory the free propagator of the massless gluon field embodying the quark-quark interaction was proportional to

$$\lambda^2/k^4, \tag{1}$$

where λ is a constant with the dimensions of mass and k is the gluon four-momentum. As a result the Schwinger mechanism[4] manifestly occurred, and it was shown that any charged particle was totally screened by vacuum polarization effects. In addition, explicit calculations of the deep-inelastic electroproduction structure functions in perturbation theory were in agreement with Bjorken scaling with corrections of $O(q^{-1})$, where q is the virtual photon four-momentum. These features of the Abelian model will also be shown to be true in the non-Abelian version. In addition, we shall argue that the quarks can be nonrelativistic within hadrons and that the spectrum of states has linearly rising Regge trajectories.

In spite of these salutary properties an interaction of the form of Eq. (1) could be questioned because of well-known[5] indefinite-metric difficulties which result in the violation of unitarity. While an optimist may hope that the nonappearance of colored gluons in asymptotic (color singlet) states might eliminate unitarity problems it is almost certain that the approximation techniques which will necessarily be used to find the bound states will lead to the occurrence of negative-metric states. Whether these states are "real" or artifacts of the approximation will not be clear. In view of this we suggested[3] that the gluon propagator be taken in principal value rather than as a Feynman propagator:

$$\mathrm{P}\frac{\lambda^2}{k^4}=\frac{\lambda^2}{2}\left[\frac{1}{(k^2+i\epsilon)^2}+\frac{1}{(k^2-i\epsilon)^2}\right]. \tag{2}$$

As a result unitarity is maintained order by order in perturbation theory. Gluons do not appear in asymptotic states. All components of the vector-gluon propagator are "Coulombized" and the gluon field reduced to the embodiment of a direct quark interaction. There are a number of other decided advantages to principal-value propagators in the present context: (1) no color singlet states composed solely of gluons, (2) the elimination of substantial infrared divergences, (3) the suppression of corrections to Bjorken scaling in the electroproduction structure functions by a factor of q^2

vis-à-vis the corresponding Feynman-propagator result which sets the stage for precocious scaling, and (4) the elimination of closed loops of vector gluons and thus the elimination of Faddeev-Popov ghost loops.

In Sec. II we give a brief recapitulation of the Abelian model. In Sec. III we describe the canonical properties of the non-Abelian model. In Sec. IV we describe the qualitative features of the model and describe an approximation technique which appears to be naturally adopted to "solving" the theory. We shall restrict our discussion to the color binding interaction and defer the introduction of other interactions to a later work. The properties of the bound states in the non-Abelian model are currently under study and will be the subject of the next report.

II. ABELIAN MODEL

The possibility that the physical particle spectrum of a field theory consisted only of neutral states and did not include states of charged fields was first investigated in massless two-dimensional quantum electrodynamics.[6] In that case the absence of the "electron" from the gauge-invariant physical particle spectrum was directly related to the acquisition of a mass by the photon via the Schwinger mechanism. The Schwinger mechanism was manifest in the lowest-order contribution to the vacuum polarization (Fig. 1), and, taking account of the dimensionality of the coupling constant, $e \sim$ mass, could almost be considered a consequence of dimensional analysis. These vacuum polarization effects led to the total screening of the "electronic" charge, and, as a result, the "electron" was removed from the gauge-invariant physical particle spectrum. Our Abelian and non-Abelian models will display a similar pattern of events.

The Lagrangian of the Abelian model contains two gluon fields, $A^1_\mu(x)$ and $A^2_\mu(x)$, and the quark field $\psi(x)$:

$$\mathcal{L} = -\tfrac{1}{2}F^1_{\mu\nu}F^2_{\mu\nu} - \tfrac{1}{2}\lambda^2 A^2_\mu A^2_\mu + \bar\psi(i\not\nabla - g\not A^1 - m)\psi\,, \tag{3}$$

where for typographic convenience we denote the inner product of four vectors, $a\cdot b = a_\mu b_\mu = a_0 b_0 - \vec a\cdot\vec b$ throughout, λ is a constant with the dimensions of mass, g is dimensionless, and $F^i_{\mu\nu} = \partial_\nu A^i_\mu - \partial_\mu A^i_\nu$.

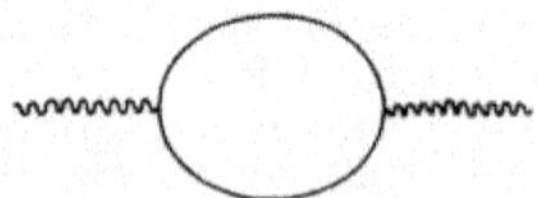

FIG. 1. A vacuum polarization diagram.

Following the canonical procedure we find the equations of motion:

$$\partial_\mu F^1_{\mu\nu} + \lambda^2 A^2_\nu = 0\,, \tag{4}$$

$$\partial_\mu F^2_{\mu\nu} + gJ_\nu = 0\,, \tag{5}$$

$$(i\not\nabla - g\not A^1 - m)\psi = 0\,, \tag{6}$$

and nonzero equal-time commutation relations [in the Coulomb gauge $\vec\nabla\cdot\vec A^1 = 0$; note $\partial_\mu A^2_\mu = 0$ by Eq. (4)]

$$[F^1_{0i}(x), A^2_j(y)] = i\Delta^{tr}_{ij}(x-y)\,, \tag{7}$$

$$[F^2_{0i}(x), A^1_j(y)] = i\Delta^{tr}_{ij}(x-y)\,, \tag{8}$$

with $i, j = 1, 2, 3$ and

$$\Delta^{tr}_{ij}(x-y) = \int \frac{d^3k}{(2\pi)^3}\, e^{i\vec k\cdot(\vec x - \vec y)}\left(\delta_{ij} - \frac{k_i k_j}{|\vec k|^2}\right). \tag{9}$$

It is clear from the equations of motion, Eqs. (4) and (5), tnat A^2_μ may be eliminated to obtain

$$\Box\partial_\mu F^1_{\mu\nu} + g\lambda^2 J_\nu = 0\,. \tag{10}$$

The form of the quark-gluon interaction and Eq. (10) show that only the Green's function of A^1_μ is relevant to quark-quark scattering. The perturbation theory rules of QED may be used if the photon propagator is replaced with the gluon propagator for A^1_μ:

$$iG^{11}_{\mu\nu}(k) = \frac{i\lambda^2(g_{\mu\nu} - \chi k_\mu k_\nu/k^2)}{k^4}\,, \tag{11}$$

where χ is constant, determined by the gauge choice.

In Ref. 3 we showed that choosing $G^{11}_{\mu\nu}$ to be a principal-value propagator allowed us to develop a perturbation theory which was unitary order by order:

$$G^{11}_{\mu\nu}(k^2) = \tfrac{1}{2}[G^{11}_{\mu\nu}(k^2 + i\epsilon) + G^{11}_{\mu\nu}(k^2 - i\epsilon)]\,. \tag{12}$$

In addition, the equivalent of the Nambu representation of a Feynman diagram was given and some features of the perturbation theory discussed. Of particular interest was a calculation of the deep-inelastic electroproduction structure functions which scaled in the Bjorken limit. Leading corrections to scaling were of $O(q^{-4})$ as $q^2 \to \infty$ with q being the virtual photon four-momentum, and were given by the diagrams of Fig. 2(b), 2(c), and 2(d). This is to be contrasted with the logarithmic deviations from scaling found in pseudoscalar or vector meson models previously studied.[7]

The Schwinger mechanism manifestly occurred in low orders of perturbation theory. As a result quarks (and all charged objects) are removed from the gauge-invariant spectrum of physical states. The total screening of charge can be seen from the following argument.[3] Consider a spatially bounded system of charge density ρ. The total charge is

$$Q = \int d^3x\, \rho(x) \tag{13}$$

$$= \frac{-1}{g\lambda^2} \int d^3x \,\Box \nabla^2 A_0^1 \tag{14}$$

using the equations of motion in the Coulomb gauge. By Gauss's law

$$Q = \frac{-1}{g\lambda^2} \int d\vec{S} \cdot \vec{\nabla} \Box A_0^1 . \tag{15}$$

From the definition of a Green's function, we have

$$A_0^1(x) = \int d^4y\, G_{00}^{11}(x-y)\rho(y) \tag{16}$$

in the Coulomb gauge. If, for simplicity, we choose ρ to describe a static point quark charge and use the free gluon propagator [Eq. (11)], then $Q \neq 0$. However, if we take account of the effect of vacuum polarization processes (the Schwinger mechanism) we find $A_0^1(x)$ is a monotonically decreasing function of $|\vec{x}|$ for large $|\vec{x}|$ and consequently $Q = 0$ in the limit where the integration surface is taken to infinity in Eq. (15). Thus the spectrum of physical states does not include states of nonzero charge. In the next section we shall show that the proof of quark confinement is essentially the same in the non-Abelian model.

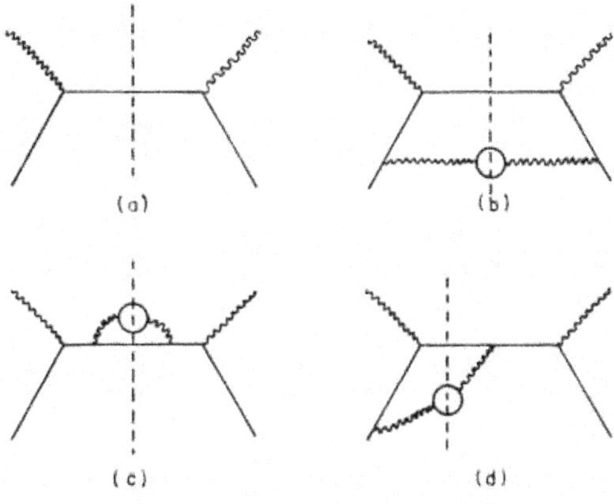

FIG. 2. Lowest-order diagrams contributing to the inelastic electroproduction structure functions. The dashed lines indicate the only contributions to the electroproduction structure functions of the absorptive part of the forward virtual Compton scattering diagram. External "wiggly" lines represent photons while internal "wiggly" lines represent gluons.

III. NON-ABELIAN MODEL

The non-Abelian model for the color sector of hadronic interactions is a direct generalization of the model of the last section.[8] There are two colored Yang-Mills fields, $A_{\mu a}^1(x)$ and $A_{\mu a}^2(x)$, which when regarded as vectors in the adjoint representation of the color group are denoted $\underline{A}_\mu^1$ and $\underline{A}_\mu^2$. The Lagrangian is

$$\begin{aligned}\mathcal{L} = {}& \tfrac{1}{2}\underline{F}_{\mu\nu}^1 \cdot \underline{F}_{\mu\nu}^2 - \tfrac{1}{2}\underline{F}_{\mu\nu}^2 \cdot (\partial_\mu \underline{A}_\nu^1 - \partial_\nu \underline{A}_\mu^1 + g\underline{A}_\mu^1 \times \underline{A}_\nu^1) \\ & - \tfrac{1}{2}\underline{F}_{\mu\nu}^1 \cdot (\partial_\mu \underline{A}_\nu^2 - \partial_\nu \underline{A}_\mu^2 + g\underline{A}_\mu^1 \times \underline{A}_\nu^2 - g\underline{A}_\nu^1 \times \underline{A}_\mu^2) \\ & - \tfrac{1}{2}\lambda^2 \underline{A}_\mu^2 \cdot \underline{A}_\mu^2 + \bar{\psi}(i\nabla\!\!\!/ + gA\!\!\!/^1 - m)\psi \end{aligned} \tag{17}$$

$$= \mathcal{L}_0 + \bar{\psi}(i\nabla\!\!\!/ + gA\!\!\!/^1 - m)\psi , \tag{18}$$

with ψ being the quark field.

It is invariant under the local gauge transformation

$$\psi' = S^{-1}\psi , \tag{19}$$

$$A_\mu^{1\prime} = S^{-1}A_\mu^1 S + \frac{i}{g} S^{-1}\partial_\mu S , \tag{20}$$

$$A_\mu^{2\prime} = S^{-1}A_\mu^2 S , \tag{21}$$

$$F_{\mu\nu}^{1\prime} = S^{-1}F_{\mu\nu}^1 S , \tag{22}$$

$$F_{\mu\nu}^{2\prime} = S^{-1}F_{\mu\nu}^2 S , \tag{23}$$

where S is an element in the gauge group G [which is color SU(3) in our case], and A_μ^1 is a matrix in the defining representation of G formed from

$$A_\mu^1 = \underline{A}_\mu^1 \cdot \underline{T} . \tag{24}$$

T_a is a matrix in the defining representation of G satisfying

$$[T_a, T_b] = i f_{abc} T_c , \tag{25}$$

and $\underline{T}$ is a vector formed from such matrices. We note that the homogeneity of the gauge transformation of A_μ^2 allows a mass term to occur in $\mathcal{L}$ without breaking the gauge symmetry. We shall see that the natural gauge-fixing term to add to the Lagrangian has the form

$$-\frac{1}{\beta}\partial_\mu \underline{A}_\mu^1 \cdot \partial_\nu \underline{A}_\nu^2 . \tag{26}$$

The Euler-Lagrange equations of motion are obtained in the canonical manner:

$$(\partial_\mu + g\underline{A}_\mu^1 \times)\underline{F}_{\mu\nu}^1 - \lambda^2 \underline{A}_\nu^2 = 0 , \tag{27}$$

$$(\partial_\mu + g\underline{A}_\mu^1 \times)\underline{F}_{\mu\nu}^2 + g\underline{A}_\mu^2 \times \underline{F}_{\mu\nu}^1 + g\underline{J}_\nu = 0 , \tag{28}$$

$$\underline{F}_{\mu\nu}^1 = \partial_\mu \underline{A}_\nu^1 - \partial_\nu \underline{A}_\mu^1 + g\underline{A}_\mu^1 \times \underline{A}_\nu^1 , \tag{29}$$

$$\underline{F}_{\mu\nu}^2 = \partial_\mu \underline{A}_\nu^2 - \partial_\nu \underline{A}_\mu^2 + g\underline{A}_\mu^1 \times \underline{A}_\nu^2 - g\underline{A}_\nu^1 \times \underline{A}_\mu^2 , \tag{30}$$

$$(i\nabla\!\!\!/ + gA\!\!\!/^1 - m)\psi = 0 . \tag{31}$$

The antisymmetry of $\underline{F}_{\mu\nu}^1$ and $\underline{F}_{\mu\nu}^2$ leads to two conservation laws,

$$\partial_\nu(g\underline{A}^1_\mu \times \underline{F}^1_{\mu\nu} - \lambda^2 \underline{A}^2_\nu) = 0\,, \tag{32}$$

$$\partial_\nu(\underline{A}^1_\mu \times \underline{F}^2_{\mu\nu} + \underline{A}^2_\mu \times \underline{F}^1_{\mu\nu} + \underline{J}_\nu) = 0\,, \tag{33}$$

which can be reexpressed as

$$(\partial_\nu + g\,\underline{A}^1_\nu \times)\underline{A}^2_\nu = 0 \tag{34}$$

and

$$(\partial_\nu + g\underline{A}^1_\nu \times)\underline{J}_\nu = 0 \tag{35}$$

using the equations of motion. The first of these relations acts in effect as a gauge-fixing term for A^2_μ if a gauge is chosen for A^1_μ. The second relation has the familiar form of current-conservation equations in conventional Yang-Mills theories.

We turn now to the derivation of the perturbation-theory rules in the gluon sector. We consider the vacuum-vacuum transition amplitude in the presence of external sources[9]:

$$W(\underline{J}^1_\mu, \underline{J}^2_\mu) = \int \prod_x dA^1_\mu\, dA^2_\mu \exp\left[i \int d^4x \left(\mathcal{L}_0 - \frac{1}{\beta}\partial_\mu \underline{A}^1_\mu \cdot \partial_\nu \underline{A}^2_\nu + \underline{A}^1_\mu \cdot \underline{J}^1_\mu + \underline{A}^2_\mu \cdot \underline{J}^2_\mu\right)\right]. \tag{36}$$

After some functional translations we find

$$W(\underline{J}^1_\mu, \underline{J}^2_\nu) \equiv \exp\left\{-i\int d^4x\, d^4y \left[\underline{J}^1_\mu(x)\cdot G^{12}_{\mu\nu}(x-y)\cdot \underline{J}^2_\nu(y) + \tfrac{1}{2}\underline{J}^1_\mu(x)\cdot G^{11}_{\mu\nu}(x-y)\cdot \underline{J}^1_\nu(y)\right]\right\}, \tag{37}$$

where we have dropped an irrelevant factor independent of $\underline{J}^1_\mu$ and $\underline{J}^2_\mu$ on the right-hand side, and

$$G^{12}_{\mu\nu ab}(x) = -\delta_{ab}\int \frac{d^4k\, e^{-ik\cdot x}}{(2\pi)^4 k^2}\left[g_{\mu\nu} + (\beta - 1)\frac{k_\mu k_\nu}{k^2}\right] \tag{38}$$

and

$$G^{11}_{\mu\nu ab}(x) = \frac{\lambda^2 \delta_{ab}}{(2\pi)^4}\int \frac{d^4k\, e^{-ik\cdot x}}{k^4}\left[g_{\mu\nu} + (\beta^2 - 1)\frac{k_\mu k_\nu}{k^2}\right], \tag{39}$$

with a and b labeling color indices. The free propagators corresponding to the time-ordered products are

$$\langle TA^1_{\mu a}(x)A^1_{\nu b}(y)\rangle = iG^{11}_{\mu\nu ab}(x-y) \tag{40}$$

and

$$\langle TA^1_{\mu a}(x)A^2_{\nu b}(y)\rangle = iG^{12}_{\mu\nu ab}(x-y)\,. \tag{41}$$

The somewhat unusual Green's functions of Eqs. (40) and (41) have their origin in the canonical equal-time commutation relations which we shall now find.

From Eqs. (27)–(30) we obtain the equations of motion

$$\partial_0 \underline{A}^1_k = \underline{F}^1_{0k} + \partial_k \underline{A}^1_0 + g\underline{A}^1_k \times \underline{A}^1_0\,, \tag{42}$$

$$\partial_0 \underline{A}^2_k = \underline{F}^2_{0k} + \partial_k \underline{A}^2_0 + g\underline{A}^2_k \times \underline{A}^1_0 - g\underline{A}^2_0 \times \underline{A}^1_k\,, \tag{43}$$

$$\partial_0 \underline{F}^1_{0k} = (\partial_i + g\underline{A}_i \times)\underline{F}^1_{ik} - g\underline{A}^1_0 \times \underline{F}^1_{0k} + \lambda^2 \underline{A}^2_k\,, \tag{44}$$

$$\partial_0 \underline{F}^2_{0k} = (\partial_i + g\underline{A}^1_i \times)\underline{F}^2_{ik} - g\underline{A}^1_0 \times \underline{F}^2_{0k} - g\underline{A}^2_\mu \times \underline{F}^1_{\mu k} - g\underline{J}_k\,, \tag{45}$$

and equations of constraint

$$\underline{F}^1_{ik} = \partial_i \underline{A}^1_k - \partial_k \underline{A}^1_i + g\underline{A}^1_i \times \underline{A}^1_k\,, \tag{46}$$

$$\underline{F}^2_{ik} = \partial_i \underline{A}^2_k - \partial_k \underline{A}^2_i + g\underline{A}^1_i \times \underline{A}^2_k - g\underline{A}^1_k \times \underline{A}^2_i\,, \tag{47}$$

$$(\partial_i + g\underline{A}^1_i \times)\underline{F}^1_{i0} + \lambda^2 \underline{A}^2_0 = 0\,, \tag{48}$$

$$(\partial_i + g\underline{A}^1_i \times)\underline{F}^2_{i0} + g\underline{A}^2_i \times \underline{F}^1_{i0} - g\underline{J}_0 = 0\,. \tag{49}$$

The Lagrangian indicates that the canonical momenta are

$$\underline{\Pi}^1_j = \underline{F}^2_{0j} \tag{50}$$

and

$$\underline{\Pi}^2_j = \underline{F}^1_{0j}\,, \tag{51}$$

for $j = 1, 2, 3$ with $\underline{\Pi}^i_j$ conjugate to $\underline{A}^i_j$, and $\underline{A}^i_0$ having no conjugate momentum for $i = 1, 2$. However, the equations of constraint indicate that not all components are independent. We now find the independent components. Let us define

$$\underline{F}^a_{0i} = \underline{F}^{aT}_{0i} + \underline{F}^{aL}_{0i} \tag{52}$$

and

$$\underline{F}^{aL}_{0i} = \partial_i \underline{\varphi}^a\,, \tag{53}$$

where

$$\partial_i \underline{F}^{aT}_{0i} = 0\,. \tag{54}$$

Then Eq. (48) gives

$$(\partial_i + g\underline{A}^1_i \times)\partial_i \underline{\varphi}^1 - \lambda^2 \underline{A}^2_0 = -g\underline{A}^1_i \times \underline{F}^{1T}_{i0} \tag{55}$$

and Eq. (49) gives

$$(\partial_i + g\underline{A}^1_i \times)\partial_i \underline{\varphi}^2 + g\underline{A}^2_i \times \partial_i \underline{\varphi}^1 = g\underline{A}^1_i \times \underline{F}^{2T}_{i0} + g\underline{A}^2_i \times \underline{F}^{1T}_{0i} - g\underline{J}_0\,. \tag{56}$$

Rewriting Eqs. (42) and (43) after taking the divergence with respect to spatial components gives

$$(\partial_0 + g\underline{A}^1_0 \times)\partial_k \underline{A}^1_k = (\partial_k + g\underline{A}^1_k \times)\partial_k \underline{A}^1_0 + \partial_k\partial_k \underline{\varphi}^1 \tag{57}$$

and

$$(\partial_0 + g\underline{A}^1_0 \times)\partial_k \underline{A}^2_k + g\underline{A}^2_0 \times \partial_k \underline{A}^1_k = \partial_k\partial_k \underline{A}^2_0 + g\underline{A}^2_k \times \partial_k \underline{A}^1_0 + g\underline{A}^1_k \times \partial_k \underline{A}^2_0 + \partial_k\partial_k \underline{\varphi}^2 \tag{58}$$

If we choose the Coulomb gauge, $\vec{\nabla}\cdot\underline{A}^1=0$, then

$$\partial_k\partial_k\underline{A}_0^1+g\underline{A}_k^1\times\partial_k\underline{A}_0^1+\partial_k\partial_k\underline{\phi}^1=0 \tag{59}$$

and

$$(\partial_0+g\underline{A}_0^1\times)\partial_k\underline{A}_k^2-\partial_k\partial_k\underline{A}_0^2-g\underline{A}_k^2\times\partial_k\underline{A}_0^1-g\underline{A}_k^1\times\partial_k\underline{A}_0^2 -\partial_k\partial_k\underline{\phi}^2=0\,, \tag{60}$$

thus determining $\underline{A}_0^1$ and $\underline{A}_0^2$. Suppose we now define

$$\vec{\underline{A}}^2=\vec{\underline{A}}^{2T}+\vec{\underline{A}}^{2L}\,, \tag{61}$$

$$\vec{\underline{A}}^{2L}=\vec{\nabla}\underline{\phi}^3\,, \tag{62}$$

with

$$\vec{\nabla}\cdot\vec{\underline{A}}^{2T}=0 \tag{63}$$

Taking the divergence of Eq. (44) leads to our final equation for dependent variables

$$\lambda^2\partial_k\partial_k\underline{\phi}^3=\partial_0\partial_k\partial_k\underline{\phi}^1+g\partial_k(\underline{A}_\mu^1\times\underline{F}_{\mu k}^1)\,. \tag{64}$$

The independent dynamical variables are thus seen to be F_{0i}^{1T}, F_{0i}^{2T}, A_i^{1T}, and A_i^{2T}. Their equal-time commutation relations are

$$[F_{0ia}^{1T}(x),A_{jb}^2(y)]=i\delta_{ab}\Delta_{ij}^{tr}(x-y)\,, \tag{65}$$

$$[F_{0ia}^{2T}(x),A_{jb}^1(y)]=i\delta_{ab}\Delta_{ij}^{tr}(x-y)\,, \tag{66}$$

with $i,j=1,2,3$, Δ_{ij}^{tr} given by Eq. (9), and a and b are color indices. All other commutators of the forms $[A^1,A^1]$, $[A^2,A^2]$, $[F^1,F^1]$, $[F^2,F^2]$, $[F^1,F^2]$ are zero.

We return to our development of perturbation-theory rules. The cubic and quartic gluon vertices of our model are given by (see Fig. 3)

$i\Gamma_{\lambda\mu\nu}^{abc}(p,q,r)\equiv$

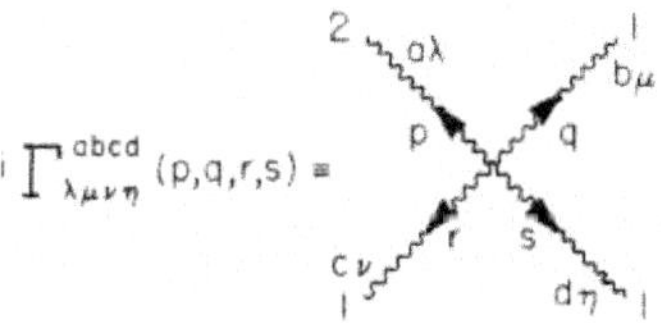

FIG. 3. Cubic and quartic vertices which are given in Eqs. (67) and (68). They introduce $1/r$ potentials in the model and may have an important effect in the baryon spectrum. The numbers 1 and 2 indicate fields $\underline{A}_\mu^1$ and $\underline{A}_\mu^2$, respectively, while p, q, r, and s are momenta, and a, b, c, and d are color indices.

$$i\Gamma_{\lambda\mu\nu}^{abc}(p,q,r)=f^{abc}[g_{\nu\lambda}(r_\mu-p_\mu)+g_{\mu\lambda}(p_\nu-q_\nu) +g_{\mu\nu}(q_\lambda-r_\lambda)]\,, \tag{67}$$

with $p+q+r=0$, and

$$\begin{aligned}i\Gamma_{\lambda\mu\nu\eta}^{abcd}(p,q,r,s)=&-if^{abf}f^{cdf}(g_{\lambda\nu}g_{\mu\eta}-g_{\eta\mu}g_{\lambda\nu})\\&-if^{acf}f^{bdf}(g_{\lambda\eta}g_{\nu\mu}-g_{\nu\lambda}g_{\mu\eta})\\&-if^{adf}f^{bcf}(g_{\lambda\eta}g_{\nu\mu}-g_{\lambda\mu}g_{\nu\eta})\,,\end{aligned} \tag{68}$$

with $p+q+r+s=0$.

The Faddeev-Popov ghost loops will not be relevant to our line of development so we omit their discussion. The necessity for their introduction[10] is closely related to the requirement of unitarity in Yang-Mills theories. In the present model unitarity will be necessarily violated irrespective of the ghost loops if the Green's functions [Eqs. (38) and (39)] pole ambiguities are resolved by using Feynman's $i\epsilon$ procedure. To avoid unitarity violation we have suggested an alternative procedure where the Green's function singularities are taken in principal value,

$$G_{\mu\nu ab}^{kL}(k^2)=\tfrac{1}{2}[G_{\mu\nu ab}^{kL}(k^2+i\epsilon)+G_{\mu\nu ab}^{kL}(k^2-i\epsilon)]\,, \tag{69}$$

in momentum space (cf. the Appendix). This choice has the advantage stated in the Introduction. The effects are the same as in the Abelian model[3] and may be summarized as: (1) Only states composed solely of quarks contribute to unitarity sums, (2) gluons do not appear in asymptotic states, (3) unitarity is achieved but at the price of possible advanced effects whose range is limited to hadronic dimensions and thus apparently unobservable, and (4) nonscaling corrections to Bjorken scaling in the deep-inelastic electroproduction structure functions are suppressed by a factor of q^2 vis-à-vis the corresponding result using Feynman propagators with q being the virtual photon four-momentum.

A novel feature of the use of principal-value propagators in non-Abelian models is the elimination of closed loops composed solely of gluons. If we consider a subdiagram consisting of a gluon loop with p lines, then Eq. (51) of Ref. 3 gives the Feynman parameter representation

$$I=\int_{-\infty}^{\infty}\prod_{j=1}^{p}\alpha_j\,d\alpha_j\,\frac{\epsilon(\alpha_1\alpha_2\cdots\alpha_p C)}{C^2}\,N e^{iD/C}\,, \tag{70}$$

where C is a polynomial consisting of Feynman parameters only, while D contains scalar products of external momenta, N symbolizes appropriate numerator factors, and $\epsilon(\alpha)=\pm1$ if $\alpha\gtrless0$. Since N can be written as a sum of terms each of which is homogeneous in the Feynman parameters, we can take N to be homogeneous without loss of generality. Then scaling all parameters with u, assuming

$$N(u\alpha_1, u\alpha_2, \ldots, u\alpha_p) = u^r N(\alpha_1, \alpha_2, \ldots, \alpha_p), \quad (71)$$

with r an integer, and using

$$\int_0^\infty \frac{du}{u}\,\delta\left(1 - \frac{|\alpha_1 + \alpha_2 + \cdots + \alpha_p|}{u}\right) = 1 \quad (72)$$

we find

$$I = \Gamma(r + 2p - 2L) \times \int_{-\infty}^{\infty} \frac{\prod_{j=1}^{p} \alpha_j\, d\alpha_j \epsilon(\alpha_1 \alpha_2 \cdots \alpha_p C) N \delta\left(1 - \left|\sum_k \alpha_k\right|\right)}{C^2(-iD/C)^{r+2p-2L}}, \quad (73)$$

with L = number of loops = 1. Suppose we let $\alpha_j \to -\alpha_j$ for all j in I. Then we find $I = -I$ or

$$I = 0. \quad (74)$$

Thus any closed loop containing only principal-value propagators is zero. Since Faddeev-Popov ghosts appear only in closed loops and consistency[11] requires we use principal-value propagators for them if we use such propagators for gluons, we see that ghosts do not appear in our model. Physically we can understand this result if we remember that ghost loops were introduced to cure problems arising from contributions to unitarity sums of "opened" gluon loops.[10] In our model "opened" loops do not contribute to unitarity sums in any case so the raison d'être for ghosts is lacking.

We now derive the Ward-Takahashi-Slavnov identities using functional methods. Since we take our gluon propagators in principal value it might appear that our use of functional techniques is unjustified. We shall take the view that the functional representation of the vacuum-vacuum transition amplitude embodies the combinatorics of perturbation theory and acts as a generating function for identities, such as the Ward-Takahashi-Slavnov identities. Thus, questions of convergence of functional integrals are irrelevant—the important question is whether identities are valid in perturbation theory.

We define $W(J)$, the vacuum-vacuum transition amplitude, by

$$W(J) = \int \prod_x dA_\mu^1\, dA_\mu^2\, d\psi\, d\bar{\psi} \exp\left(i \int \tilde{\mathfrak{L}}\, dx\right), \quad (75)$$

with

$$\tilde{\mathfrak{L}} = \mathfrak{L} - \frac{1}{\beta}\partial_\mu \underline{A}_\mu^1 \cdot \partial_\nu \underline{A}_\nu^2 + \underline{A}_\mu^1 \cdot \underline{J}_\mu^1 + \underline{A}_\mu^2 \cdot \underline{J}_\mu^2 + \bar{\psi}\eta + \bar{\eta}\psi,$$

with $\mathfrak{L}$ given by Eq. (17). Under the infinitesimal gauge variation

$$\underline{A}_\mu^1 \to \underline{A}_\mu^1 - (\partial_\mu + g\underline{A}_\mu^1 \times)\underline{\theta}, \quad (76)$$

$$\underline{A}_\mu^2 \to \underline{A}_\mu^2 - g\underline{A}_\mu^2 \times \underline{\theta}, \quad (77)$$

$$\psi \to \psi - ig\theta\psi, \quad (78)$$

$$\bar{\psi} \to \bar{\psi} + ig\bar{\psi}\theta, \quad (79)$$

with $\theta = \underline{T} \cdot \underline{\theta}$, $\mathfrak{L}$ is invariant but the remaining terms in $\tilde{\mathfrak{L}}$ lead to

$$\begin{aligned}\delta\tilde{\mathfrak{L}} = \frac{1}{\beta}[(\partial_\mu + g\underline{A}_\mu^1 \times)\partial_\nu \partial_\mu \underline{A}_\nu^2 + g\underline{A}_\nu^2 \times \partial_\mu \partial_\nu \underline{A}_\mu^1] \cdot \underline{\theta} \\ - (\partial_\mu + g\underline{A}_\mu^1 \times)\underline{J}_\mu^1 \cdot \underline{\theta} - g\underline{J}_\mu^2 \times \underline{A}_\mu^2 \cdot \underline{\theta} \\ + ig\bar{\psi}\theta\eta - ig\bar{\eta}\theta\psi. \end{aligned} \quad (80)$$

Since a transformation of the integration variables does not change the value of the functional integral, the variation of W with respect to θ can be taken to be zero and our equivalent of the Ward-Takahashi-Slavnov identity is

$$\left\{\frac{1}{\beta}\left[D_\nu\left(\frac{\delta}{i\delta \underline{J}_\sigma^1}\right)\partial_\nu \partial_\mu \frac{\delta}{i\delta \underline{J}_\mu^2} + g\frac{\delta}{i\delta \underline{J}_\nu^2} \times \partial_\nu \partial_\mu \frac{\delta}{i\delta \underline{J}_\mu^1}\right] + D_\mu\left(\frac{\delta}{i\delta \underline{J}_\alpha^1}\right)\underline{J}_\mu^1 - g\underline{J}_\mu^2 \times \frac{\delta}{i\delta \underline{J}_\mu^2} + g\underline{T}\eta\frac{\delta}{\delta\eta} - g\bar{\eta}\underline{T}\frac{\delta}{\delta\bar{\eta}}\right\} W = 0, \quad (81)$$

with

$$D_\mu\left(\frac{\delta}{i\delta \underline{J}_\sigma^1}\right) = \partial_\mu + g\frac{\delta}{i\delta \underline{J}_\mu^1} \times. \quad (82)$$

In order to investigate the structure of the gluon propagators we shall obtain the proper vertex identity equivalent to Eq. (81). We focus on the novelties of the gluon sector and neglect the quark field terms in $\tilde{\mathfrak{L}}$ and Eq. (81). Let us define

$$W(J) = e^{iZ(J)}, \quad (83)$$

$$\underline{B}_\mu^i = -\frac{\delta Z(J)}{\delta \underline{J}_\mu^i}, \quad i = 1, 2 \quad (84)$$

$$\Gamma(B) = Z(J) + \int d^4x(\underline{J}_\mu^1 \cdot \underline{B}_\mu^1 + \underline{J}_\mu^2 \cdot \underline{B}_\mu^2), \quad (85)$$

where $\Gamma(B)$ is the generating functional of proper vertices. An immediate consequence is

$$\underline{J}_\mu^i = \frac{\delta\Gamma}{\delta \underline{B}_\mu^i}, \quad i = 1,2 \tag{86}$$

and as a result Eq. (81) can be rewritten in the form

$$\frac{1}{\beta}\left[\Box\partial_\mu \underline{B}_\mu^2 - g\underline{B}_\nu^1 \times \partial_\nu\partial_\mu \underline{B}_\mu^2 - g\underline{B}_\nu^2\times\partial_\mu\partial_\nu \underline{B}_\mu^1 + g\frac{\delta}{i\delta \underline{J}_\nu^1}\times\partial_\nu\partial_\mu \underline{B}_\mu^2 + g\frac{\delta}{i\delta \underline{J}_\nu^2}\times\partial_\nu\partial_\mu \underline{B}_\mu^1\right] - \partial_\mu\frac{\delta\Gamma}{\delta \underline{B}_\mu^1} + \underline{B}_\mu^1\times\frac{\delta\Gamma}{\delta \underline{B}_\mu^1} + \underline{B}_\mu^2\times\frac{\delta\Gamma}{\delta \underline{B}_\mu^2} = 0. \tag{87}$$

If we apply $\delta/\delta \underline{B}_\alpha^i$ to Eq. (87) and set $\underline{B}_\mu^i = 0$ afterwards, we find

$$-\partial_\mu \left.\frac{\delta^2\Gamma}{\delta \underline{B}_\alpha^1 \delta \underline{B}_\mu^1}\right|_{\underline{B}^1 = \underline{B}^2 = 0} = 0. \tag{88}$$

The second-order functional derivative of Γ is the inverse of the full propagator $G_{\mu\nu ab}^{11\prime}$ and Eq. (88) implies that the proper part of $(G_{\mu\nu ab}^{11\prime})^{-1}$ is purely transverse. We note that the "free" propagator (Eq. 39) contribution to $(G_{\mu\nu ab}^{11\prime})^{-1}$ is not one-particle irreducible and thus not constrained by Eq. (88). Therefore we find the general form

$$G_{\mu\nu ab}^{11\prime}(k) = \delta_{ab}\left(g_{\mu\nu} - \frac{k_\mu k_\nu}{k^2}\right)G^{11}(k^2) + \delta_{ab}\beta^2\lambda^2\frac{k_\mu k_\nu}{k^6}, \tag{89}$$

so that the longitudinal part of the full propagator is not renormalized.

The longitudinal part of the full propagator $G_{\mu\nu ab}^{12\prime}(k)$ is also not renormalized. This may be seen by applying $\delta/\delta \underline{B}_\mu^2$ to Eq. (87) and setting $\underline{B}_\mu^i = 0$ afterwards:

$$\frac{1}{\beta}\Box\partial_\mu\delta^4(x-y) - \partial_\nu \left.\frac{\delta^2\Gamma}{\delta \underline{B}_\mu^2 \delta \underline{B}_\nu^1}\right|_{B^1 = B^2 = 0} = 0. \tag{90}$$

This implies

$$(G_{\mu\nu ab}^{12\prime})^{-1} = \frac{\delta_{ab}(g_{\mu\nu} - k_\mu k_\nu/k^2)}{G^{12}} - \frac{k_\mu k_\nu \delta_{ab}}{\beta} \tag{91}$$

or

$$G_{\mu\nu ab}^{12\prime}(k) = \delta_{ab}\left(g_{\mu\nu} - \frac{k_\mu k_\nu}{k^2}\right)G^{12}(k) - \beta\delta_{ab}\frac{k_\mu k_\nu}{k^4}. \tag{92}$$

Having now developed the general form of the propagators we now will define the gluon vacuum polarization tensors,

$$\Pi_{\mu\nu ab}^{11}(k) = [G_{\mu\nu ab}^{11\prime}(k)]^{-1} - [G_{\mu\nu ab}^{11}(k)]^{-1}, \tag{93}$$

$$\Pi_{\mu\nu ab}^{12}(k) = [G_{\mu\nu ab}^{12\prime}(k)]^{-1} - [G_{\mu\nu ab}^{12}(k)]^{-1}, \tag{94}$$

which are transverse by our previous discussion:

$$k_\mu\Pi_{\mu\nu ab}^{11} = k_\mu\Pi_{\mu\nu ab}^{12} = 0. \tag{95}$$

Rather than write the Schwinger-Dyson equations for our polarization tensors we have given a diagrammatic representation in Fig. 4.

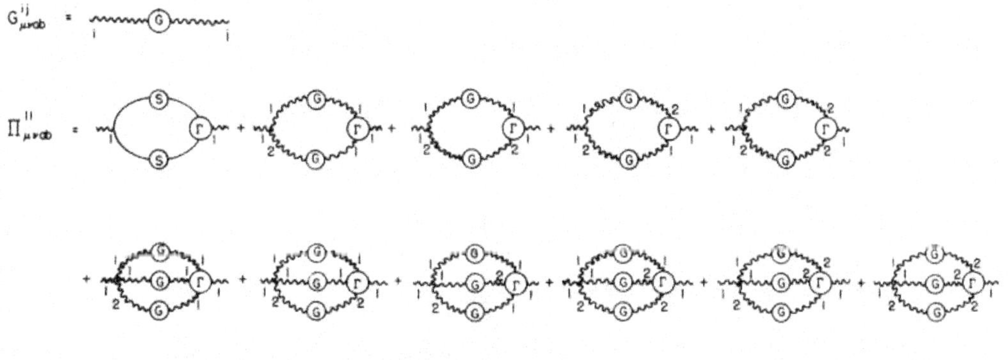

FIG. 4. Diagrammatic representation of the Schwinger-Dyson equation for the proper gluon self-energy, $\Pi_{\mu\nu ab}^{11}$. The numbers at the end of a gluon line specify whether $\underline{A}_\mu^1$ or $\underline{A}_\mu^2$ correspond to that end. The quark propagator is denoted S while Γ denotes the appropriate proper (one-particle irreducible) vertex function. A similar diagrammatic expression can be written for $\Pi_{\mu\nu ab}^{12}$.

IV. OBSERVATIONS

The Schwinger mechanism forces quark confinement to bound color singlet states in a manner which is identical to the Abelian case as described in Sec. II. In order to demonstrate that only color singlets exist in the gauge-invariant physical particle spectrum it is sufficient to show

$$\underline{Q}\psi_{\text{phys}} = 0\,, \tag{96}$$

where

$$\underline{Q} = \int d^3x\, \underline{J}_0(x) \tag{97}$$

for any physical state ψ_{phys} corresponding to a spatially localized distribution of quarks. We consider a single static quark located at the origin and choose to work in the Coulomb gauge ($\vec{\nabla}\cdot\vec{A}^1 = 0$). Then the time components of the equations of motion [Eqs. (27) and (28)] lead to (at large distance)

$$\Box\nabla^2 \underline{A}_0^1 = g\lambda^2 \underline{J}_0 \tag{98}$$

if we take into account the elimination of gluons' degrees of freedom through the choice of principal-value propagators and their consequent inability to act as sources. We may now repeat the arguments of Eqs. (13)–(16) for the Abelian case after noticing the occurrence of the Schwinger mechanism in the non-Abelian case which can be verified in low orders of perturbation theory for $\Pi^{11}_{\mu\nu ab}$. Thus the expectation value of the charge in the one-quark state is zero. Since the one-quark state is a charge eigenstate, we find Eq. (96) to be true in this case and more generally through the additivity of the charge operator. Thus only color singlet bound states of quarks are physical.[12]

While the infrared behavior of the theory leads to quark confinement, the ultraviolet behavior allows the quarks to appear quasifree. This is particularly noticeable when we take $\lambda^2 = 0$ in our Lagrangian and examine the corresponding perturbation theory. Taking $\lambda^2 = 0$ is equivalent to examining the short-distance behavior of the theory. The only diagrams which exist in this limit are given in Fig. 5. The quark sector of the theory is free. The only nontree structures are one-quark-loop diagrams for the scattering of gluons associated with A_μ^2 (which of course can only be generated by a hypothetical external source). (As a point of comparison we have shown in Fig. 6 the additional diagrams which would occur in the even that Feynman propagators were used—these diagrams necessarily involve gluon loops which principal-value propagators force to be zero.) The vital role of the $\lambda^2 A_\mu^2 A_\mu^2$ term in the Lagrangian in generating the interacting theory and the fact that λ^2 has the dimensions of (mass)^2 allow a natural approximation procedure in this model. This is perhaps best seen within the context of deep-inelastic electroproduction. Just as in the Abelian case we find that the structure functions scale with leading corrections of $O(q^{-4})$, where q equals the virtual photon four-momentum. We can establish a parton picture of scattering wherein the photon is absorbed on one of the quasifree nucleon constituents [as in Fig. 2(a)] if $|q^2| \gg g^2\lambda^2$. Then leading corrections to such a picture [e.g., the diagrams of Fig. 2(b)–2(d)] will be suppressed by $(g^2\lambda^2/q^2)^2$. Thus the dimensional nature of the effective coupling constant allows a particularly simple picture to exist of the region of large spacelike virtual photon mass and the parton picture emerges as a natural approximation.

The k^{-4} form of the quark interaction also appears to have decidedly good features as far as the bound-state structure is concerned. Ignoring the numerator tensor (which does not affect our conclusions), we find the Fourier transform of the gluon propagator,

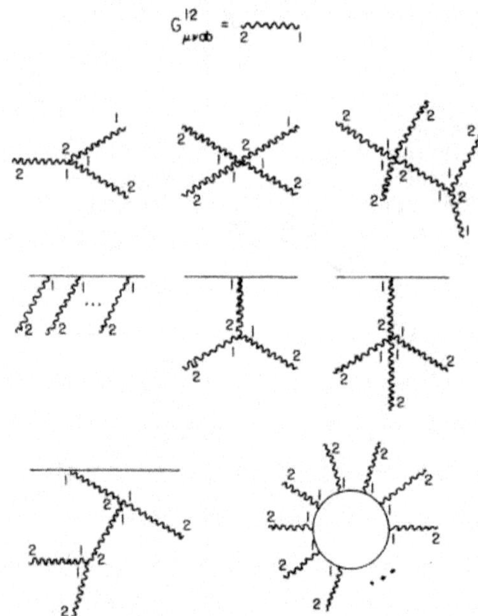

FIG. 5. Some examples of the surviving diagrams in the $\lambda^2 = 0$ limit of the non-Abelian model with principal-value gluon propagators. Except for the class of one-fermion-loop diagrams only tree diagrams exist in this limit. Note that there are no four (or more) external quark line diagrams and no two (or more) external $\underline{A}_\mu^1$ gluon "external" lines.

$$G(k) = \mathrm{P}\frac{1}{k^4}\,, \tag{99}$$

to be proportional to

$$\bar{G}(x) = \theta(x^2)\,. \tag{100}$$

Since $\bar{G}$ has a smooth finite limit as $x^2 \to 0$, the short-distance limit, arguments can be made[13] that low-mass bound states can occur in this model. In addition, Dalitz[14] has pointed out that the linearity of trajectories on the Chew-Frautschi plot would follow from a flat-bottomed, smooth interaction—a criterion which is met by Eq. (100). [It is interesting to note that had we used a Feynman propagator rather than principal value, then $\bar{G}$ would have been $\ln(x^2)$ and thus the general criterion just stated would not have been met. This would appear to be another point in favor of our choice of principal-value propagators.]

Another property which is desirable in the bound-state solutions is nonrelativistic motion of the bound-state constituents.[15] Again an interaction of the form of Eq. (99) appears to realize this feature—even in the strong-binding limit. To see this we shall first take account of the Schwinger mechanism and in the spirit of Hartree-Fock theory modify the quark interaction to

$$G'(k) = \mathrm{P}\frac{1}{(k^2 - \mu^2)^2}\,. \tag{101}$$

If we now take Eq. (101) to be the Green's function for the effective gluon field and calculate the "Coulomb potential" of a static, point quark source located at the origin we find

$$\varphi(r) = \frac{\varphi_0}{\mu}\, e^{-\mu r}\,, \tag{102}$$

where ϕ_0 is a constant independent of μ. In the limit $\mu \to 0$ we find

$$\phi(r) \cong \varphi_0\left(\frac{1}{\mu} - r + \cdots\right). \tag{103}$$

The first two terms of Eq. (103) correspond to choosing Eq. (99) rather than Eq. (101) as the gluon Green's function (in the limit $\mu \to 0$). Equation (102) includes vacuum polarization effects which damp the interaction at large distances. Thus Eq. (102) imperfectly reflects the possibility that a quark-antiquark pair can separate and induce another quark-antiquark pair to be created from the vacuum so that two color singlet mesons will result (presuming it is energetically favored). At shorter distances Eq. (102) appears to be a reasonable approximation. This exponential potential was studied within the framework of the Schrödinger equation in the strong-binding limit (ϕ_0/μ large) by Greenberg.[16] He showed that the average momentum of the bound constituent in the s state satisfied

$$\frac{p}{m} \sim \left(\frac{\mu}{m}\right)^{1/3} \tag{104}$$

with m being the quark mass. Thus for μ/m small the quark motion is self-consistently nonrelativistic.

In conclusion, we have shown that a four-dimensional, Lorentz-invariant second-quantized field theory of hadron binding is possible with scaling electroproduction structure functions, only zero-triality physical particle states, and, apparently, linearly rising Regge trajectories and nonrelativistic constituents. A detailed study of the bound states is now in progress.

ACKNOWLEDGMENT

I am grateful to the members of the Newman Laboratory for interesting conversations.

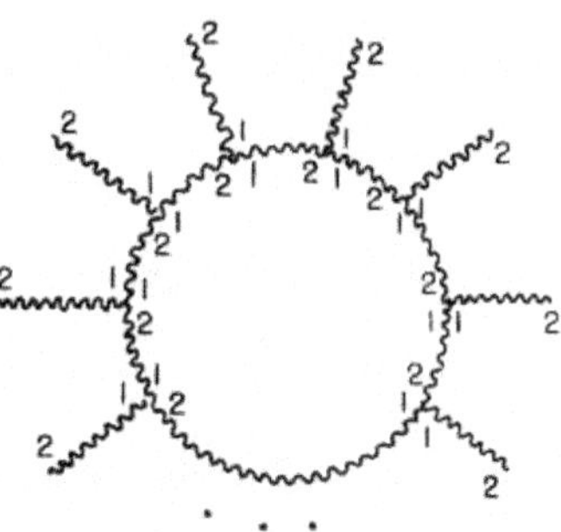

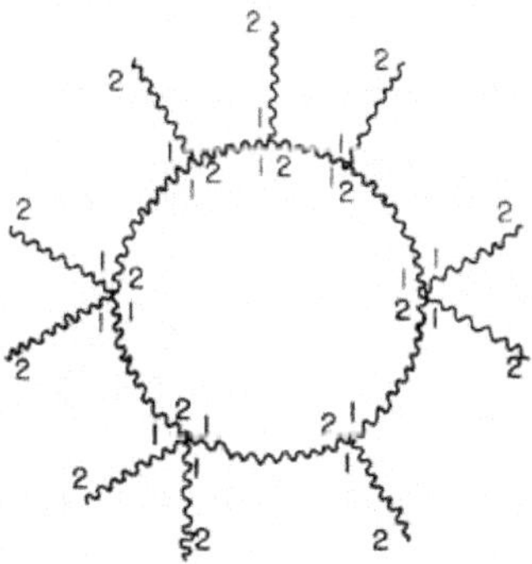

FIG. 6. Some additional diagrams which occur in the $\lambda^2 = 0$ limit of the non-Abelian model if Feynman gluon propagators are used. In addition, there will be Faddeev-Popov ghost-loop diagrams depending on the choice of gauge.

APPENDIX

In Ref. 3 semiclassical arguments based on Dirac's theory of constraints were given to introduce the use of principal-value propagators. We will now describe a second-quantized realization of those arguments for the case of a scalar Klein-Gordon field $\varphi(x)$ with the Lagrangian

$$\mathcal{L} = \tfrac{1}{2}(\partial_\mu \varphi)^2 - \tfrac{1}{2}m^2\varphi^2 . \tag{A1}$$

The generalization to vector gluons is immediate. The canonical equal-time commutation relations are

$$[\phi, \varphi] = [\dot{\phi}, \dot{\phi}] = 0 , \tag{A2}$$

$$[\dot{\phi}(\vec{x}, t), \varphi(\vec{y}, t)] = -i\delta^3(\vec{x} - \vec{y}) . \tag{A3}$$

If we expand $\phi(x)$ in plane waves,

$$\varphi(\vec{x}, t) = \sum_{\vec{k}} (A_{\vec{k}}^- e^{-ik\cdot x} + A_{\vec{k}}^+ e^{ik\cdot x}) , \tag{A4}$$

then the q-number Fourier components $A_{\vec{k}}^\mp$ must satisfy

$$[A_{\vec{k}}^-, A_{\vec{k}'}^-] = [A_{\vec{k}}^+, A_{\vec{k}}^+] = 0 , \tag{A5}$$

$$[A_{\vec{k}}^-, A_{\vec{k}'}^+] = \delta^3(\vec{k}' - \vec{k}) \tag{A6}$$

for consistency with Eqs. (A2) and (A3). Now the time-ordered product satisfies

$$T(\varphi(x)\varphi(y)) = \epsilon(x_0 - y_0)[\varphi(x), \phi(y)] + \{\varphi(x), \varphi(y)\} , \tag{A7}$$

with $\epsilon(x_0) = \pm 1$ for $x_0 \gtrless 0$ and $\{A, B\} = AB + BA$. The first term on the right-hand side is a c number completely determined by Eqs. (A5) and (A6). If the second q-number expression were zero, then we would obtain a principal-value propagator from Eq. (A7):

$$T(\varphi(x)\phi(y)) = i\int \frac{d^4k}{(2\pi)^4} e^{-ik\cdot(x-y)}\, \mathrm{P}\,\frac{1}{(k^2 - m^2)} . \tag{A8}$$

We therefore require

$$\{\phi(x), \varphi(y)\} = 0 , \tag{A9}$$

with the consequence

$$\{A_{\vec{k}}^-, A_{\vec{k}'}^-\} = \{A_{\vec{k}}^+, A_{\vec{k}'}^+\} = \{A_{\vec{k}}^-, A_{\vec{k}'}^+\} = 0 . \tag{A10}$$

Equations (A5), (A6), and (A10) imply

$$A_{\vec{k}}^- A_{\vec{k}'}^- = A_{\vec{k}}^+ A_{\vec{k}'}^+ = 0 , \tag{A11}$$

$$A_{\vec{k}}^- A_{\vec{k}'}^+ = \tfrac{1}{2}\delta^3(\vec{k} - \vec{k}') , \tag{A12}$$

$$A_{\vec{k}}^+ A_{\vec{k}'}^- = -\tfrac{1}{2}\delta^3(\vec{k} - \vec{k}') \tag{A13}$$

for all $\vec{k}$ and $\vec{k}'$. Thus quadratic terms in A and A^+ are reduced to c numbers. It should further be noted that the multiplication rule is not associative.[17] In fact, the multiplication rules of the A and A^+ operators in Eqs. (A11)–(A13) are realized by taking multiplication to be

$$UV = \tfrac{1}{2}[U, V] \tag{A14}$$

for U, V being any $A_{\vec{k}}^-$ or $A_{\vec{k}'}^+$. If we take an analogy to Lie-algebra theory seriously, where the adjoint representation of an algebra has a multiplication rule defined by commutators

$$\tilde{U} * \tilde{V} = [\tilde{U}, \tilde{V}] \tag{A15}$$

then we could call Eqs. (A11)–(A13) the adjoint representation of the Fourier components of φ.

The c-number nature of AA, A^+A^+, or AA^+ can be understood physically in the following manner. Since the ϕ field has principal-value propagators it is not associated with a particle but is merely the embodiment of an interaction between other objects (which we have suppressed in our Lagrangian). Consequently an emission of a ϕ field quantum must be directly correlated with a subsequent absorption—it cannot propagate into empty space. The c-number nature of AA^+ reflects this correlation between emission and absorption.

Finally, it should be noted that the existence of a vacuum is inconsistent with Eqs. (A11)–(A13).

---

*Work supported in part by the National Science Foundation.

[1] K. Johnson, Phys. Rev. D 6, 1101 (1972); C. M. Bender, J. E. Mandula, and G. S. Guralnik, Phys. Rev. Lett. 32, 1467 (1974); A. Chodos *et al.*, Phys. Rev. D 9, 3471 (1974); W. A. Bardeen *et al.*, *ibid.* 11, 1094 (1975); M. Creutz, *ibid.* 10, 1749 (1974); P. Vinciarelli, Nuovo Cimento Lett. 4, 905 (1972); R. Dashen, B. Hasslacher, and A. Neveu, Phys. Rev. D 10, 4114 (1974); 10, 4130 (1974); 10, 4138 (1974).

[2] Y. Nambu, in ***Preludes in Theoretical Physics***, edited by A. de-Shalit, H. Feshbach, and L. Van Hove (North-Holland, Amsterdam, 1966), p. 133; H. J. Lipkin, Phys. Lett. 45B, 267 (1973).

[3] S. Blaha, Phys. Rev. D 10, 4268 (1974).

[4] J. Schwinger, Phys. Rev. 128, 2425 (1962).

[5] A. Pais and G. Uhlenbeck, Phys. Rev. 79, 145 (1950); J. Kiskis, Phys. Rev. D 11, 2178 (1975).

[6] A. Casher, J. Kogut, and L. Susskind, Phys. Rev. D 10, 732 (1974); J. Lowenstein and J. Swieca, Ann. Phys. (N.Y.) 68, 172 (1971).

[7] R. Jackiw and G. Preparata, Phys. Rev. Lett. 22, 975 (1969); S. Adler and W. Tung, *ibid.* 22, 978 (1969); S. Blaha, Phys. Rev. D 3, 510 (1971).

[8]The Lagrangian of Eq. (17) was first written by D. Sinclair as a generalization of the Abelian model of Ref. 3. An alternative non-Abelian model for quark confinement has been suggested by S. K. Kauffmann [Nucl. Phys. B87, 133 (1975)]. I am grateful to Dr. Kauffmann for sending me a copy of his paper prior to publication.

[9]E. Abers and B. W. Lee [Phys. Rep. 9C, 1 (1973)] provide a useful review of conventional Yang-Mills theories.

[10] R. P. Feynman, Acta Phys. Pol. 24, 697 (1963).

[11]B. W. Lee and J. Zinn-Justin [Phys. Rev. D 5, 3121 (1972)] point out that the $i\epsilon$ prescription in their Eq. (2.8) for the ghost loop is dictated by unitarity considerations.

[12]This does not preclude color-singlet states of the gluons from playing a role in the theory. They are not particles but can be exchanged between color-singlet quark states in scattering events. On naive dimensional grounds they should be most important in forward scattering. This leads to the possibility that the Pomeron might possibly be interpreted as a "two-gluon bound state". In the case of wide-angle scattering the predominant mechanism for large momentum transfer would appear to be constituent interchange due to the strong damping effects of k^{-4} propagators on momentum transfer.

[13]M. Böhm, H. Joos, and M. Krammer, in *Recent Developments in Mathematical Physics*, proceedings of the XII Schladming Conference (Acta Phys. Austriaca Suppl. XI), edited by P. Urban (Springer, New York, 1970), p. 3.

[14]R. H. Dalitz, a paper presented at the Topical Conference on Meson Spectroscopy, Philadelphia, 1968 (unpublished).

[15]H. J. Lipkin, Phys. Rep. 8C, 175 (1973).

[16]O. W. Greenberg, Phys. Rev. 147, 1077 (1966).

[17]Nonassociative field operators have been previously used by M. Günaydin and F. Gürsey, Phys. Rev. D 9, 3387 (1974).

Appendix B. Abelian Quark Confinement

This refereed paper is S. Blaha, Phys. Rev. D**10**, 4268 (July, 1974). Reprinted with the kind permission of Physical Review D.

Landau-Ginzburg theory, but $\rho \sim i\langle\phi^*\dot{\phi} - \dot{\phi}^*\phi\rangle = 0$ in the Higgs theory.

[9]M. Kalb and P. Ramond, Phys. Rev. D 9, 2273 (1974). See also E. Cremmer and J. Scherk, Nucl. Phys. B72, 117 (1974).

[10]L. N. Chang and F. Mansouri, Phys. Rev. D 5, 2235 (1972); Goto, Ref. 7; G. Goddard, J. Goldstone, C. Rebbi, and C. B. Thorn, Nucl. Phys. B56, 109 (1973).

[11]A clearcut answer to this problem seems to be lacking. See, however, Nielsen and Olesen, Ref. 1; G. 't Hooft, CERN Report No. TH-1873-CERN, 1974 (unpublished); Y. Nambu, Ref. 2.

PHYSICAL REVIEW D VOLUME 10, NUMBER 12 15 DECEMBER 1974

Towards a field theory of hadron binding*

Stephen Blaha

Laboratory of Nuclear Studies, Cornell University, Ithaca, New York 14850

(Received 17 July 1974)

A field-theoretic model for hadron binding is described in which free quarks are totally screened. Quarks interact via a dipole vector-gluon field. A second-quantization procedure for the gluon field, which reduces the field to an embodiment of a direct particle interaction, eliminates unitarity problems. A detailed description of perturbation-theory rules is given. In contrast to the results of the pseudoscalar-meson and massive-vector-meson models (without cutoff), scaling occurs in the electroproduction structure functions. Another possible model having some resemblance to the relativistic harmonic-oscillator quark model of Feynman, Kislinger, and Ravndal is also described. It is unitary and has scaling structure functions.

I. INTRODUCTION

The current understanding of hadronic structure allows two apparently contradictory statements to be made: The constituents of the hadron appear to be loosely bound, quasifree particles. The constituents of the hadron are not produced and do not occur outside of hadrons. Several attempts have been made to resolve this paradoxical situation. They may be divided into two categories: "conventional" field-theoretic approaches,[1,2] and *ad hoc* approaches which postulate manifestly nonfield-theoretic structures for confinement, e.g., the "bag" model.[3] In the first approach Casher, Kogut, and Susskind[1] and Wilson[2] showed that quarks could be totally screened and not observed. However, a four-dimensional, Lorentz-invariant field-theoretic model of hadron binding with its attendant conceptual and computational advantages appears to be lacking. We shall discuss a possible candidate, the dipole gluon model, in detail. In addition, another possibility is briefly described in Appendix B which bears some comparison with the quark model of Feynman *et al.*[4] The dipole gluon model has two major qualitative features in common with the bag model[3] and the two-dimensional quantum-electrodynamic model[1]: (1) The dipole gluon field has no independent degrees of freedom; neither does a bag or the two-dimensional electromagnetic field. (2) The "Coulomb" potential between quarks is proportional to the distance between them in all three models. In a sense the bag model may be regarded as a phenomenological approximation to the dipole model, and the dipole model as a generalization of the two-dimensional model to four dimensions.

In Sec. II we describe a quantization procedure which avoids the introduction of indefinite-metric in or out states and thus leads to a unitary S matrix. In Sec. III we describe the properties of the "free" gluon Lagrangian model. In Sec. IV we describe the perturbation-theory rules of the dipole model. Section V contains a discussion of unitarity, causality, quark confinement, and scaling properties of the electroproduction structure functions. For simplicity we shall ignore all but the dipole quark interaction and do not introduce internal quark quantum numbers.

II. SECOND-QUANTIZATION PROCEDURE FOR THE GLUON FIELD

We shall not quantize the gluon field in the conventional manner for three reasons: (1) to be consistent with experiment where no such particle has been identified, (2) to avoid unitarity problems in the S matrix, and (3) to avoid infrared problems in perturbation theory. We attribute no dynamical degrees of freedom to the gluon field. Instead we regard the field as the embodiment of a direct quark-quark interaction. The gluon field can thus be removed from the Lagrangian in favor of a nonlocal current-current interaction. However, it will be of no small technical advantage to keep the gluon field in the Lagrangian. In order to do this we shall second-quantize the field following the normal prescription and then, instead of introducing a Fock space for free gluons, reduce q-number expressions in the gluon field to c-number expressions via suitable operator boundary conditions.

To illustrate this procedure we consider a scalar boson field, ϕ, with Lagrangian L. We second-quantize the in field, ϕ_{in}, with equal-time commutators

$$[\phi_{in}(x), \phi_{in}(y)] = 0\,, \tag{1}$$

$$[\Pi_{in}(x), \Pi_{in}(y)] = 0\,, \tag{2}$$

$$[\phi_{in}(x), \Pi_{in}(y)] = -i\delta^3(\vec{x}-\vec{y})\,, \tag{3}$$

where

$$\Pi_{in}(x) = \frac{\delta L}{\delta\dot{\phi}_{in}(x)} \tag{4}$$

and L_F is the free Lagrangian part of L. The usual operator expressions and identities are established. In particular the formal expansion of the S matrix in terms of time-ordered products of in fields can be made. (We are using only in fields for convenience—our remarks apply to out fields also.)

The unequal-time commutator, $[\phi_{in}(x), \phi_{in}(y)]$, is a c-number expression which is completely determined if we require that it be consistent with the equations of motion, that it be consistent with Eqs. (1)–(3) in the limit of equal times, and that it vanish at spacelike distances. Consequently all terms with an even number of factors of $\phi(x)$ reduce to sums of c numbers times products of anticommutators $\{\phi(x), \phi(y)\}$. Terms with an odd number of factors have one factor, $\phi(x)$, times sums of c numbers times products of anticommutators. At this point we could introduce a Fock space of states to complete the reduction of q-number expressions to c number expressions. For reasons stated above we do not. In analogy to Dirac's theory[5] of Hamiltonian constraints we impose operator boundary conditions which complete the specification of the dynamics of the system. We choose

$$\Pi_{in}(x) \approx 0 \approx \phi_{in}(x) \tag{5}$$

for all x, where $\approx$ means weakly equal in the sense of Dirac, i.e., evaluate all commutators before imposing the constraints. This eliminates ϕ's degrees of freedom. The free Hamiltonian,

$$H_F = \int \Pi\dot{\phi} - L_F\,, \tag{6}$$

is now arbitrary to the extent that H_F may be replaced by

$$H_T = H_F + \int A\phi_{in} + \int b\Pi_{in}\,, \tag{7}$$

where A and b will be completely determined by requiring Eq. (5) be true for all time:

$$[\Pi_{in}, H_T] \approx 0 \tag{8}$$

and

$$[\phi_{in}, H_T] \approx 0\,. \tag{9}$$

Thus

$$A = -\frac{\delta H_F}{\delta\phi_{in}} \tag{10}$$

and

$$b = -\frac{\delta H_F}{\delta\Pi_{in}}\,. \tag{11}$$

To see the effects of this procedure more concretely let

$$L_F = \int (\tfrac{1}{2}\partial_\mu\phi\partial^\mu\phi - \tfrac{1}{2}m^2\phi^2)\,; \tag{12}$$

then (suppressing the subscript "in" for notational convenience)

$$i\Delta(x-y) = [\phi(x), \phi(y)] \tag{13}$$

$$= \int \frac{d^4k}{(2\pi)^3}\,\epsilon(k_0)\delta(k^2-m^2)e^{-ik\cdot(x-y)} \tag{14}$$

and the time-ordered product becomes

$$i\bar{\Delta}(x-y) \equiv T(\phi(x)\phi(y)) = \tfrac{1}{2}i\epsilon(x_0-y_0)\Delta(x-y)\,, \tag{15}$$

with $\epsilon(x) = \pm 1$ for $x \gtrless 0$. More generally, for even N

$$T(\phi(1)\phi(2)\cdots\phi(N))$$

$$= \sum_{\text{permutations}} i^{N/2}\bar{\Delta}(x_1-x_2)\bar{\Delta}(x_3-x_4)\cdots\bar{\Delta}(x_{N-1}-x_N)\,, \tag{16}$$

where $\phi(i) = \phi(x_i)$. The natural correspondence to the Wick expansion

$$\langle 0|T(\psi(x_1)\cdots\psi(x_N))|0\rangle$$

$$= \sum_{\text{permutations}} i^{N/2}\Delta_F(x_1-x_2)\cdots\Delta_F(x_{N-1}-x_N) \tag{17}$$

(where Δ_F is the Feynman propagator corresponding to the field ψ) allows us to use conventional perturbation-theory rules, except that diagrams with incoming or outgoing ϕ lines do not contribute to the S matrix and the Feynman propagator

$$\Delta_F(k) = \frac{1}{k^2 - m^2 + i\epsilon} \tag{18}$$

is to be replaced with

$$\bar{\Delta}(k) = \mathrm{P}\,\frac{1}{k^2 - m^2 + i\epsilon} = \frac{1}{2}\left(\frac{1}{k^2 - m^2 + i\epsilon} + \frac{1}{k^2 - m^2 - i\epsilon}\right) \tag{19}$$

for internal ϕ lines. In configuration space the Green's function corresponding to Eq. (19) is half the sum of the advanced and retarded Green's functions.

If we follow the above procedure in second-quantizing the electromagnetic field the resulting model quantum electrodynamics corresponds to the classical action-at-a-distance electrodynamics of Schwarzschild, Tetrode, and Fokker.[6] The fact that photon production does not occur in the model QED corresponds to the absence of radiation reaction in the classical theory. In Sec. V this will be shown to be the key to maintaining the unitarity of the S matrix in the dipole gluon model.

III. THE DIPOLE GLUON MODEL

We now consider a model[4] for hadron binding which has several major qualitative features in agreement with experimental results: large-transverse-momenta damping, scaling electroproduction structure functions, and complete screening of free quarks. The Lagrangian is

$$\mathcal{L} = -\tfrac{1}{2}F^1_{\mu\nu}F^{2\mu\nu} - \tfrac{1}{2}\lambda^2 A^2_\mu A^{2\mu} + \bar{\psi}(i\not\nabla - g\not A^1 - m)\psi\,, \tag{20}$$

where A^1_μ and A^2_μ are massless gluon fields, $F^i_{\mu\nu} = \partial_\nu A^i_\mu - \partial_\mu A^i_\nu$ for $i = 1, 2$, ψ is the quark field, and g is a dimensionless and λ a dimensional coupling constant. The equations of motion are

$$\partial^\mu F^1_{\mu\nu} + \lambda^2 A^2_\nu = 0\,, \tag{21}$$

$$\partial^\mu F^2_{\mu\nu} + gJ_\nu = 0\,, \tag{22}$$

$$(i\not\nabla - g\not A^1 - m)\psi = 0\,, \tag{23}$$

with J_μ the quark current. Equation (21) implies $\partial^\mu A^2_\mu = 0$ while A^1_μ is a gauge-invariant field. As a result we have

$$\Box\,\partial^\mu F^1_{\mu\nu} + g\lambda^2 J_\nu = 0\,. \tag{24}$$

We now consider the "free" gluon case whose Lagrangian is the first two terms on the right-hand side of Eq. (20). The canonical momentum conjugate to A^1_μ is

$$\Pi^1_\mu = F^2_{0\mu} \tag{25}$$

and that conjugate to A^2_μ is

$$\Pi^2_\mu = F^1_{0\mu}\,. \tag{26}$$

Since $\Pi^1_0 = \Pi^2_0 = 0$ we find that A^1_0 and A^2_0 are c numbers and thus $\vec{\nabla}\cdot\vec{A}^2$ is also a c number with the possible exception of the zero-frequency mode. If we choose the Coulomb gauge for A^1_μ

$$\vec{\nabla}\cdot\vec{A}^1 = 0 \tag{27}$$

then we obtain the equal-time commutation relations

$$[\Pi^a_i(x), A^b_j(y)] = i\delta^{ab}\int \frac{d^3k}{(2\pi)^3}\, e^{i\vec{k}\cdot(\vec{x}-\vec{y})}\left(\delta_{ij} - \frac{k_i k_j}{|\vec{k}|^2}\right) \tag{28}$$

for $i, j = 1, 2, 3$, in analogy to similar expressions in quantum electrodynamics. All other equal-time commutators are zero. We can define an electric field, $\vec{E}$, and magnetic field, $\vec{B}$, by

$$\vec{E} = -\vec{\nabla}A^{10} - \frac{\partial}{\partial t}\vec{A}^1 \tag{29}$$

and

$$\vec{B} = \vec{\nabla}\times\vec{A}^1\,, \tag{30}$$

which imply

$$\vec{\nabla}\times\vec{E} = \frac{\partial\vec{B}}{\partial t} \tag{31}$$

and

$$\vec{\nabla}\cdot\vec{B} = 0\,. \tag{32}$$

In the Coulomb gauge Eq. (24) can be restated as

$$\Box\vec{\nabla}\cdot\vec{E} = g\lambda^2 J^0\,, \tag{33}$$

$$\Box\left(\vec{\nabla}\times\vec{B} - \frac{\partial\vec{E}}{\partial t}\right) = g\lambda^2\vec{J}\,, \tag{34}$$

where J^μ is the quark current. Equations (29) and (33) give our analog to the differential equation for the instantaneous Coulomb potential of QED,

$$\Box\nabla^2 A^{10} = -g\lambda^2 J^0\,, \tag{35}$$

while the equivalent vector-potential differential equation is

$$\Box(\Box\vec{A}^1 + \vec{\nabla}\dot{A}^{10}) = g\lambda^2\vec{J}\,. \tag{36}$$

The free gluon unequal-time commutators may be determined from Eqs. (21), (22), (28), (35), and (36) (with the current, of course, set to zero):

$$i\Delta_{ij}^{11}(x-y)\equiv[A_i^1(x),A_j^1(y)]=-i\lambda^2(\delta_{ij}-\nabla_i\nabla_j\nabla^{-2})\left[\frac{\partial}{\partial\mu^2}\Delta(x-y,\mu)\right]_{\mu=0}, \tag{37}$$

$$i\Delta_{ij}^{12}(x-y)\equiv[A_i^1(x),A_j^2(y)]=i(\delta_{ij}-\nabla_i\nabla_j\nabla^{-2})\Delta(x-y,0), \tag{38}$$

$$i\Delta_{ij}^{22}(x-y)\equiv[A_i^2(x),A_j^2(y)]=0, \tag{39}$$

with $i,j=1,2,3$ and

$$i\Delta(x-y,\mu)=\int\frac{d^4k}{(2\pi)^3}\epsilon(k_0)\delta(k^2-\mu^2)e^{-ik\cdot(x-y)}. \tag{40}$$

The commutators, Δ^{11} and Δ^{12}, are zero at spacelike separations due to the form of Δ. They are consistent with the equal-time commutation relations in that limit and they are also consistent with the equations of motion due to the identity

$$\Box\left[\frac{\partial}{\partial\mu^2}\Delta(x-y,\mu)\right]_{\mu=0}=-\Delta(x-y,0). \tag{41}$$

Assuming that we have established all operator expressions we are now in a position to apply operator boundary conditions to the gluon field. The key quantities so far as the perturbation theory we will consider in the next section is concerned are the time-ordered propagators of the gluon field

$$T(A_i^a(x)A_j^b(y))\equiv\tfrac{1}{2}\epsilon(x_0-y_0)[A_i^a(x),A_j^b(y)] \tag{42}$$

$$=\tfrac{1}{2}i\epsilon(x_0-y_0)\Delta_{ij}^{ab}(x-y), \tag{43}$$

where we have suppressed the "in" subscript on the field operator. We can take advantage of the gauge invariance of A_μ^1 to express $T(A_\mu^1A_\nu^1)$ in the Feynman gauge,

$$T(A_\mu^1(x)A_\nu^1(y))=i\lambda^2g_{\mu\nu}\int\frac{d^4k}{(2\pi)^4}\left(\mathrm{P}\frac{1}{k^4}\right)e^{-ik\cdot(x-y)} \tag{44}$$

with

$$\mathrm{P}\frac{1}{k^4}\equiv\frac{1}{2}\left[\frac{1}{(k^2+i\epsilon)^2}+\frac{1}{(k^2-i\epsilon)^2}\right]. \tag{45}$$

In coordinate space

$$T(A_\mu^1(x)A_\nu^1(y))=ig_{\mu\nu}\lambda^2\theta((x-y)^2)/16\pi. \tag{46}$$

The equations of motion of the dipole model display a close analogy to those of quantum electrodynamics. The main difference (with important physical consequences) is the increased degree of the differential equation for A_μ^1 vis-â-vis the corresponding QED equations. The result is a dipole propagator rather than a monopole propagator in momentum space. One could have second-quantized the dipole field in a manner which leads to dipole Feynman propagators. In that case the S matrix would not be unitary in perturbation theory.

IV. PERTURBATION THEORY

The rules for forming the integral corresponding to a Feynman diagram in the dipole model are identical with those of quantum electrodynamics[7] except that we use

$$iB_{\mu\nu}(q)=ig_{\mu\nu}\lambda^2\mathrm{P}\frac{1}{q^4} \tag{47}$$

rather than the Feynman photon propagator

$$iD_{F\mu\nu}(q)=-\frac{ig_{\mu\nu}}{q^2+i\epsilon}. \tag{48}$$

The choice of a principal-value propagator has substantial effects in perturbation theory. For example we shall show that consistency with unitarity requires no diagrams with in or out gluon lines contribute to the S matrix. In addition, there are novelties in the type of divergences in diagrams and the analytic structure of the S matrix. It also appears that conclusions based on summing only a finite number of graphs contributing to an S-matrix element may be misleading. This follows from the fact (to be shown in Sec. V) that free quarks do not exist upon summation of all orders of perturbation theory [Eq. (65)], though this is not seen in a summation to any finite order. The physical states are neutral bound states and thus it appears that the best methods of exploring the physics embodied in this model will involve Bethe-Salpeter equations[8] or eikonal summations. They are currently under study.

We now describe the modifications necessary to compute diagrams in perturbation theory. The propagator of Eq. (47) may be exponentiated through the use of the identity

$$\mathrm{P}\frac{1}{k^4}=-\tfrac{1}{2}\int_{-\infty}^{\infty}d\alpha\,\alpha\,\epsilon(\alpha)\exp(i\alpha k^2), \tag{49}$$

where $\epsilon(\alpha)=\pm1$ for $\alpha\gtrless0$. Since Feynman parameters are not necessarily positive the following identity will be useful in evaluating loop integrations:

$$\int d^4k\exp[iC(\alpha)k^2]=i\pi^2\epsilon(C)/C^2. \tag{50}$$

As a result the Feynman parameter representation of a diagram will have the form

$$I=\int_{-\infty}^{\infty}\prod_{j=1}^{p}\alpha_j\,d\alpha_j\int_0^{\infty}\frac{d\beta_1\cdots d\beta_q}{C^2}\,\epsilon(\alpha_1\alpha_2\cdots\alpha_p C)Ne^{iD/C}\,. \tag{51}$$

where α_i corresponds to an internal gluon line and β_i to an internal fermion line, N symbolizes numerator terms, and C and D are determinantal functions.[9] If we had given the gluons dipole Feynman propagators we would have obtained

$$\int_0^{\infty}\frac{\prod_{j=1}^{p}\alpha_j\,d\alpha_j\,d\beta_1 d\beta_2\cdots d\beta_q\,N\exp(iD/C)}{C^2}\,, \tag{52}$$

in comparison to Eq. (51). If we now scale all Feynman parameters with u and use the identity

$$\int_0^{\infty}\frac{du}{u}\,\delta\left(1-\frac{|\alpha_1+\alpha_2+\cdots+\alpha_p+\beta_1+\beta_2+\cdots+\beta_q|}{u}\right)=1 \tag{53}$$

(where $|\;|$ indicates absolute value) to introduce an integration over u in Eq. (51), we obtain

$$I=\Gamma(q+2p-2l)\int_{-\infty}^{\infty}\prod_{j=1}^{p}\alpha_j\,d\alpha_j\int_0^{\infty}\frac{d\beta_1\cdots d\beta_q\,\epsilon(\alpha_1\cdots\alpha_p C)\tilde N}{C^2(-iD/C)^{q+2p-2l}}\,\delta\left(1-\left|\sum_i\alpha_i+\sum_j\beta_j\right|\right) \tag{54}$$

where l = the number of loop integrations in the original diagram and $\tilde N$ is obtained from N. An example of this procedure is given in Appendix A. As an alternative to the above method one can introduce light-cone coordinates and evaluate pole terms by contour integrations with Eq. (45) specifying the location of the poles relative to the contour.

The divergences occurring in this model are somewhat novel. As one would expect, with a dipole propagator the ultraviolet divergences are restricted to some lower-order diagrams and are logarithmic in nature (see Fig. 1). The dipole propagator, because it is in principal value, does not induce infrared divergences in loop integrations. However, a third type of divergence, which may be called a light-cone divergence,[10] does occur and is connected with a divergence in a loop integration, $\int d^4k$, associated with the region where $k^2=k_0^2-k_3^2-\vec{k}_\perp^{\,2}\approx 0$ and $k_0, k_3\to\infty$. In the Feynman parameter representation of Eq. (54) the divergence will appear at the $\pm\infty$ limits of Feynman parameter integrals. The worst divergence is quadratic and associated with one-loop diagrams with one internal gluon line (Fig. 2). These divergences can be managed through the use of Pauli-Villars regularization. Some diagrams containing light-cone divergences are given in Fig. 2. It should be noted that they are necessarily one-loop diagrams. We can demonstrate this by an examination of the overall degree of light-cone divergence of a graph in the representation of Eq. (54). Let us scale all Feynman parameters in Eq. (54) with Λ and determine the leading behavior as $\Lambda\to\infty$. We find, for $l\equiv$ number of loops >1,

$$\tilde N\sim\Lambda^0\,, \tag{55}$$

$$C\sim\Lambda^l\,, \tag{56}$$

$$D\sim\Lambda^{l+1}\,, \tag{57}$$

and as a result

$$I\sim\Lambda^{2p+q-1-2l-(q+2p-2l)}=\Lambda^{-1}\,, \tag{58}$$

or convergence of the integral as a whole. However, for one-loop diagrams ($l=1$)

$$C\sim\Lambda^0 \tag{59}$$

and consequently

$$\tilde N\sim\Lambda^q\,, \tag{60}$$

$$I\sim\Lambda^{3-2p}\,. \tag{61}$$

For example, the diagram of Fig. 2(a) has $p=1$ and diverges quadratically.

Light-cone divergences stem directly from the use of principal-value propagators for the gluon. As such they reflect the nontrivial nature of Wick rotation in this model and they lead to divergences in the vertex renormalization constant, the wave-function renormalization constant, and the quark self-mass which prevent this model from being a superrenormalizable theory of the conventional variety.

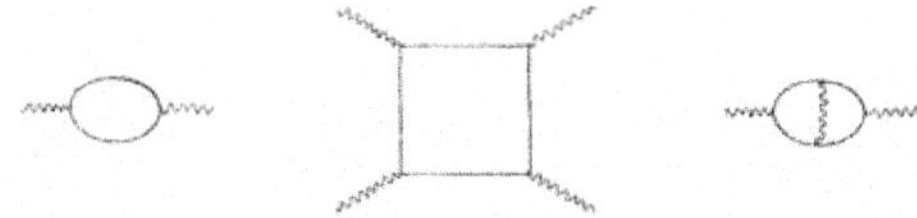

FIG. 1. Some ultraviolet-divergent diagrams.

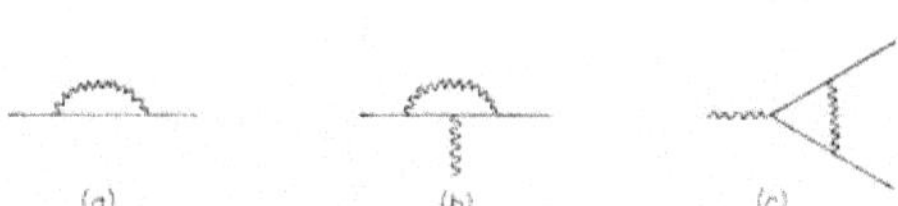

FIG. 2. Some light-cone divergent diagrams.

The Schwinger-Dyson equations for dipole electrodynamics are quite similar to those of quantum electrodynamics, with the exception of the gluon Green's functions, which we now discuss. The proper gluon self-energy, $\Pi_{\mu\nu}(q)$, which couples only to the A^1_μ channel due to the form of our Lagrangian, satisfies

$$\Pi_{\mu\nu}(q) = i\int \frac{d^4k}{(2\pi)^4}\,\mathrm{Tr}\,\gamma_\mu S'_F(k)\Gamma_\nu(k, k+q)S'_F(k+q) \tag{62}$$

$$= (q_\mu q_\nu - g_{\mu\nu}q^2)\Pi(q^2)\,, \tag{63}$$

where S'_F is the quark propagator and Γ_ν is the proper vertex function (see Fig. 3). The gluon self-energy is related to the complete gluon propagator, $B'_{\mu\nu}$, by

$$iB'_{\mu\nu} = iB_{\mu\nu} + ig^2 iB_{\mu\lambda}\Pi^{\lambda\sigma}iB'_{\sigma\nu}\,, \tag{64}$$

with $B_{\mu\nu}$ given by Eq. (47). Using Eq. (63) we find

$$B'_{\mu\nu}(q) = \frac{\lambda^2 g_{\mu\nu}}{q^4 + g^2\lambda^2 q^2\Pi(q)} - \frac{q_\mu q_\nu g^2\lambda^4\Pi(q)}{q^8 + g^2\lambda^2 q^6\Pi(q)}\,. \tag{65}$$

The Green's function for the gluon field, A^2_μ, which is zero within the context of the free gluon Lagrangian [cf. Eq. (39)], is nonzero in the interacting theory due to vacuum-polarization effects. It is related to $B'_{\mu\nu}$ by

$$B^2_{\mu\nu}(q) = \frac{(q^4/\lambda^2)B'_{\mu\nu}(q) - g_{\mu\nu}}{\lambda^2} \tag{66}$$

$$= \frac{-g^2\Pi(q)g_{\mu\nu}}{q^2 + g^2\lambda^2\Pi(q)} \tag{67}$$

up to terms proportional to $q_\mu q_\nu$. Equation (65) will play an instrumental role in the demonstration of free-quark screening in the next section.

V. SOME GENERAL PROPERTIES

In this section we will first consider the screening mechanism for quarks and then discuss unitarity, causality, and scaling properties of the lowest-order contributions to the electroproduction structure functions.

FIG. 3. Representation of the Schwinger-Dyson equation for the gluon self-energy.

In Ref. 1 attention was drawn to the screening of free quarks due to vacuum polarization. Our mechanism is a variation of the Schwinger mechanism[11] but differs from it in an important respect: It is manifest in low order and thus not a matter of conjecture—an important consideration for insoluble field theories. Even in lowest order (Fig. 1), where $\Pi(q)$ is a constant up to a logarithmic term, we find manifest screening.

Let us consider a system containing free quarks in some bounded region. We choose to work in the Coulomb gauge. Because of Eq. (35) the total charge is proportional to

$$Q \propto \int d^3x\,\nabla^2\Box A^1_0 \tag{68}$$

$$= \int d\vec{S}\cdot\vec{\nabla}\Box A^1_0\,. \tag{69}$$

However, an examination of Eq. (65) shows that important vacuum-polarization effects occur at large distances. The potential corresponding to a static free quark located at the origin is

$$A^1_0 = \frac{-\lambda^2 g|\vec{x}|}{8\pi} \tag{70}$$

(if we ignore vacuum-polarization effects) and a finite contribution to Q would result if substituted in Eq. (69). At large distances A^1_0 is substantially modified from the expression in Eq. (70). From Eq. (65) we see that the large-distance behavior of A^1_0 is controlled by

$$\frac{1}{g^2q^2\Pi(q)}\,, \tag{71}$$

and since $\Pi(q)$ is a constant up to logarithms in lowest order and not proportional to a positive power of q^2 in any finite order of perturbation theory we find A^1_0 to be proportional to at most an inverse power of $|\vec{x}|$ at large distances. Substituting an inverse power of $|\vec{x}|$ for A^1_0 in Eq. (69) and letting the surface of integration go to infinity shows $Q = 0$. Thus isolated free quarks do not exist in this model. Only neutral bound states occur.

We have chosen a propagator for the gluon which allows the S matrix to be unitarity. Our gluons are dipole ghosts, and, having indefinite metric, they would normally destroy the unitarity of the S matrix. But the quantization procedure eliminates their appearance in in or out states and their principal-value propagator precludes states containing gluons from contributing to the absorptive part of any Feynman diagram.[12] This is required if the S matrix is to be unitary. But as a result the S matrix is not analytic. The nonanalyticity is closely associated with advanced noncausal effects. Our procedure forces unitarity to be valid at the expense of noncausality. Tradeoffs of this type have recently been discussed by Coleman.[13] We return to the question of causality later.

We have verified that unitarity is maintained in perturbation theory by an explicit calculation of the lowest-order quark self-energy [Fig. 2(a)], which is

$$\Sigma(q) = \frac{-\lambda^2 g^2}{8\pi^2} \mathrm{P} \sum_i \frac{c_i \Lambda_i^{\,2} \ln(\Lambda_i^{\,2}/q^2)}{q^2(q^2 - \Lambda_i^{\,2})} \left(\frac{\Lambda_i^{\,2}}{q^2} \not{q} - 2m \right), \tag{72}$$

where $c_1 = 1$, $\Lambda_1 = m$, the regulator identities $\sum_i c_i = \sum_i c_i \Lambda_i^{\,2} = 0$ hold, and P signifies $\Sigma(q^2 + i\epsilon) = \Sigma(q^2 - i\epsilon)$ as is demonstrated in detail in Appendix A. The fact that the singularities in Eq. (72) occur in principal value implies Σ has no absorptive part. This is to be contrasted with the corresponding quantity in QED which has an absorptive part reflecting the physically allowed decay of an off-shell electron into an electron and a photon. No similar possibility exists in our model.

We now will show that the absorptive part (in the physical region) of a Feynman diagram with one internal gluon line only receives contributions from intermediate states (obtained by appropriately "cutting" internal lines in all possible ways) which do not contain the gluon. The generalization to diagrams with many gluon lines is immediate. First we note that a principal-value propagator may be decomposed:

$$\mathrm{P} \frac{1}{k^2 - m^2} = \frac{1}{k^2 - m^2 + i\epsilon} + i\pi\delta(k^2 - m^2). \tag{73}$$

For the sake of simplicity we shall write the integral corresponding to our hypothetical diagram as

$$I = \int d^4k \, \tilde{I} \, \mathrm{P} \frac{1}{k^4} \tag{74}$$

$$= \frac{\partial}{\partial \mu^2} \int d^4k \, \tilde{I} \, \mathrm{P} \frac{1}{k^2 - \mu^2} \bigg|_{\mu^2 = 0}, \tag{75}$$

where I and $\tilde{I}$ have indices and momenta appropriate to the diagram in question and the limits we have introduced engender no infrared difficulties due to the choice of a principal-value propagator. Substituting Eq. (73) into Eq. (75) we can decompose I into three Feynman integrals (actually their derivative with respect to mass, etc.),

$$I = \frac{\partial}{\partial \mu^2} \int d^4k \left[\frac{\tilde{I}}{k^2 - \mu^2 + i\epsilon} + i\pi\theta(k_0)\delta(k^2 - \mu^2) + i\pi\theta(-k_0)\delta(k^2 - \mu^2) \right]_{\mu = 0}, \tag{76}$$

in each of which only Feynman propagators are used. The last two terms correspond to opening up the loop containing the gluon. Their Feynman diagrams have in and out gluon lines of momentum k, which is summed over. Let us now restrict ourselves to the physical region[14] of our diagram so that we can take the absorptive part of I in the following way:

$$\begin{aligned} \mathrm{Abs}(I) &= i\pi \frac{\partial}{\partial \mu^2} \int d^4k [-\tilde{I}'\theta(k_0)\delta(k^2 - \mu^2) + \tilde{I}'\theta(k_0)\delta(k^2 - \mu^2) + \tilde{I}'\theta(-k_0)\delta(k^2 - \mu^2)] + R \\ &= R. \end{aligned} \tag{77}$$

The term in square brackets contains all contributions from intermediate states containing a gluon, while R contains the remainder of the absorptive part. The first two terms cancel, while the third term is zero in the physical region. Thus we have shown that the absorptive part receives no contributions from states containing the gluon. Consequently only states containing quarks contribute to unitarity sums for absorptive parts and diagrams containing external gluons do not contribute to the S matrix.

The principal-value gluon propagator has introduced noncausal effects into our model in the sense that the corresponding configuration-space Green's function is half-advanced and half-retarded. However, because we have maintained the commutativity of field operators at spacelike distances the principle of microscopic causality is not violated. Although advanced effects are not observed in everyday life they do not lead to internal inconsistency or paradoxes.[6] On the microscopic level, for example, within the confines of a hadron, advanced effects are not necessarily ruled out on physical grounds. From the earlier discussion of vacuum-polarization effects it is clear that noncausalities must be limited to very short distances. It thus appears that the only significant question involving causality is whether the nonanalyticity of the S matrix for low-order quark-quark scattering will be reflected in the scattering amplitudes of bound states in a manner which is in substantial disagreement with our understanding of S-matrix analyticity for physical particle scattering. The answer to this question is not known.

As an application of the dipole model we shall

study the deep-inelastic electroproduction structure functions in low-order perturbation theory. Previous calculations[15] of the structure functions in pseudoscalar-meson or neutral-vector-meson field-theoretic models (without transverse-momentum cutoffs) contained logarithmic deviations from scaling in apparent conflict with experimental results. The dipole model has strong transverse-momentum damping and as a result one obtains scaling structure functions—in fact, the only asymptotically leading contribution appears to be the diagram of Fig. 4(a). Higher-order diagrams do not scale by powers of q^2, the photon mass squared. For example the diagrams of lowest order in q^2 [Figs. 4(b)–4(d)] contributing to νW_2 are of $O(q^{-4})$. Thus the dipole model establishes a parton picture of the deep-inelastic structure functions since quarks appear to be pointlike particles in the scaling region. The choice of a principal-value propagator for the gluon has the effect of suppressing corrections to the scaling part of νW_2 which would have been of $O(q^{-2})$, such as the contribution of the diagram of Fig. 5. The absorptive part of that diagram is zero due to principal-value gluon propagator. Thus the precocious nature of scaling could be connected with the properties of the principal-value gluon propagator in electroproduction. On the other hand, the principal-value propagator will not play such an important role (at least in low order) in suppressing nonscaling contributions to the absorptive part of the amplitude associated with $e^+e^- \to$ hadrons. Thus low-order calculations are suggestive so far as scaling phenomena are concerned.

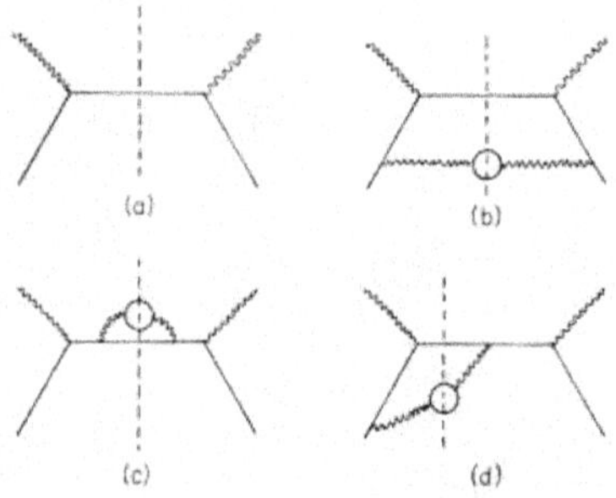

FIG. 4. Lowest-order diagrams contributing to the inelastic electroproduction structure functions. The dashed lines indicate the only contributions to the electroproduction structure functions of the absorptive part of the forward virtual Compton scattering diagram. External "wiggly" lines represent photons, while internal "wiggly" lines represent dipole gluons.

VI. CONCLUSION

The dipole electrodynamics model which we have discussed in the preceding sections is a prototype for a field theory of hadron binding. It has a number of desirable qualitative features such as quark confinement and scaling electroproduction structure functions. The physical content of this model is in the bound-states sector. This sector is currently under study using Bethe-Salpeter and eikonal techniques.

In a more realistic version of this model charge will be replaced by color in such a way that only zero-triality states are physical. The fields A_μ^1 and A_μ^2 will then become Yang-Mills fields. In that case the use of principal-value propagators appears to substantially simplify the model since closed loops of Yang-Mills fields are necessarily zero.[16]

ACKNOWLEDGMENT

I am grateful to the members of the Newman Laboratory of Nuclear Studies for helpful discussions and particularly to Dr. J. Kogut, Dr. D. K. Sinclair, and Dr. L. Susskind.

APPENDIX A

As an example of the modifications in perturbation-theory calculations resulting from principal-value propagators we evaluate the self-energy contribution of Fig. 2(a) and verify Eq. (72):

$$\Sigma(q) = \frac{ig^2\lambda^2}{(2\pi)^4}\int \frac{d^4k}{(q+k)^2 - m^2 + i\epsilon}\left(\mathrm{P}\,\frac{1}{k^4}\right)\gamma_\nu(\not{q}+\not{k}+m)\gamma^\nu \tag{A1}$$

in the Feynman gauge. Feynman parameters can be introduced, and using Eqs. (49) and (50) we obtain

$$\Sigma(q) = \frac{ig^2\lambda^2}{32\pi^2}\int_{-\infty}^{\infty} d\alpha\,\epsilon(\alpha)\alpha\int_0^{\infty}\frac{d\beta}{C^2}\left(\frac{-2\alpha}{C}\not{q}+4m\right) \times \epsilon(C)e^{iD/C}, \tag{A2}$$

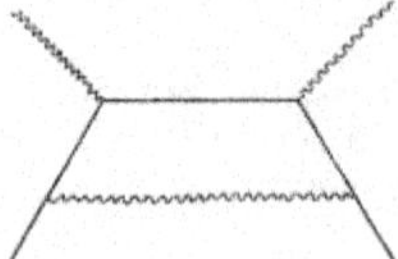

FIG. 5. A forward virtual Compton scattering diagram *not* contributing to the electroproduction structure functions.

where

$$C = \alpha + \beta \tag{A3}$$

and

$$D = \alpha\beta q^2 - \beta C m^2 \ . \tag{A4}$$

Scaling α and β and using Eq. (53) converts $\Sigma(q)$ to the form

$$\Sigma(q) = \frac{-g^2\lambda^2}{32\pi^2} \int_{-\infty}^{\infty} d\alpha\, \alpha\epsilon(\alpha) \int_0^{\infty} \frac{d\beta}{D} [-2\alpha\epsilon(C)\not{q} + 4m] \times \delta(1 - |\alpha+\beta|) \ . \tag{A5}$$

Of the two "points" contributing to the integral, $\alpha+\beta=\pm 1$, the contribution of the term $\alpha+\beta=-1$ can be included in the other term by an extension of the β integration domain:

$$\Sigma(q) = \frac{-g^2\lambda^2}{32\pi^2} \int_{-\infty}^{\infty} d\alpha\, \epsilon(\alpha)\alpha \int_{-\infty}^{\infty} d\beta \frac{\delta(1-\alpha-\beta)(-2\alpha\not{q}+4m)}{\alpha\beta q^2 - \beta m^2} \ . \tag{A6}$$

This may be shown by letting $\alpha \to -\alpha$ and $\beta \to -\beta$ in the term in question. Equation (A6) has divergences at $\alpha = \pm\infty$ after the β integration. These may be handled by introducing Pauli-Villars regulators of mass Λ_i satisfying

$$\sum_i c_i = 0 \ , \tag{A7}$$

$$\sum_i c_i \Lambda_i^{\,2} = 0 \ , \tag{A8}$$

$$c_1 = 1 \ , \tag{A9}$$

$$\Lambda_1 = m \ . \tag{A10}$$

Equation (A6) then becomes

$$\Sigma(q) = \frac{-g^2\lambda^2}{32\pi^2} \sum_i c_i \int_{-\infty}^{\infty} \frac{d\alpha\, \alpha\epsilon(\alpha)(-2\alpha\not{q}+4m)}{(1-\alpha)\{\alpha[q^2+i\epsilon(\alpha)\delta] - \Lambda_i^{\,2}\}} \ , \tag{A11}$$

which may be shown to give Eq. (72) by elementary integrations. Apparent singularities in the denominator of the integrand of Eq. (A11) do not lead to difficulties if we take account of the $i\epsilon$'s which we have suppressed. The $i\epsilon(\alpha)\delta$ term (δ is infinitesimal) shows $\Sigma(q)$ to be in principal value $[\Sigma(q^2+i\delta) = \Sigma(q^2 - i\delta)]$. It originates in the exponentiation of the principal-value propagator using Eq. (49).

APPENDIX B

We will briefly describe another possible model for hadron binding. Like the dipole model it is a member of a class of null-metric gluon theories with multipole Green's functions. The physical motivation for considering this model is a gross similarity to a quark model of Feynman *et al.*[4] which posited a relativistic harmonic oscillator potential, $x^\mu x_\mu$, between quarks and obtained quite successful agreement with experiment. If we neglect factors due to its vectorial nature (and also vacuum polarization effects) the interaction between quarks in our model is $x^\mu x_\mu \theta(x_\mu x^\mu)$ (note that $x_\mu x^\mu = x_0^{\,2} - \vec{x}^2$). The $\theta(x^\mu x_\mu)$ factor, which is necessary for unitarity to be maintained, seems to imply that only "timelike" excitations are physical, and as a result the analysis of Feynman *et al.* cannot be directly appropriated for our use.

We shall introduce three vector-gluon fields, $A_\mu^i(x)$ $(i = 1, 2, 3)$, of which only one will interact directly with the prototype spin-$\frac{1}{2}$ quark field, $\psi(x)$:

$$\mathfrak{L} = -\tfrac{1}{2}F^1_{\mu\nu}F^{3\mu\nu} + \tfrac{1}{4}F^2_{\mu\nu}F^{2\mu\nu} - \lambda^2 A^2_\mu A^{3\mu} + \bar{\psi}(i\nabla - g\not{A}^1 - m)\psi \ , \tag{B1}$$

with $F^i_{\mu\nu} = \partial_\nu A^i_\mu - \partial_\mu A^i_\nu$, and λ and g coupling constants. Following the canonical procedure we obtain the equations of motion

$$\partial^\mu F^1_{\mu\nu} + \lambda^2 A^2_\nu = 0 \ , \tag{B2}$$

$$\partial^\mu F^2_{\mu\nu} - \lambda^2 A^3_\nu = 0 \ , \tag{B3}$$

$$\partial^\mu F^3_{\mu\nu} + g J_\nu = 0 \ , \tag{B4}$$

$$(i\not{\nabla} - g\not{A}^1 - m)\psi = 0 \ , \tag{B5}$$

with J^ν the quark current. The equations of motior reveal the Lagrangian to be invariant under local gauge transformations of A^1_μ and ψ while

$$\partial^\mu A^2_\mu = \partial^\mu A^3_\mu = 0 \ . \tag{B6}$$

Furthermore, Eqs. (B2) and (B3) imply

$$\Box F^1_{\mu\nu} - \lambda^2 F^2_{\mu\nu} = 0 \ , \tag{B7}$$

$$\Box F^2_{\mu\nu} + \lambda^2 F^3_{\mu\nu} = 0 \ , \tag{B8}$$

and as a result

$$\Box^2 \partial^\mu F^1_{\mu\nu} = g\lambda^4 J_\nu \tag{B9}$$

irrespective of the gauge choice for A^1_μ.

Following the conventional procedure we find that the canonical equal-time commutation relations in the radiation gauge $(\vec{\nabla}\cdot\vec{A}^1 = 0)$ are

$$[F^a_{0i}(x), A^b_j(y)] = i h^{ab} \int \frac{d^3k}{(2\pi)^3} e^{i\vec{k}\cdot(\vec{x}-\vec{y})} \left(\delta_{ij} - \frac{k_i k_j}{|\vec{k}|^2}\right) \ , \tag{B10}$$

with $h^{13}=h^{31}=-h^{22}=1$ and all other $h^{ab}=0$. We can now choose to quantize the theory as described in the text. Again we may use the perturbation-theory rules of QED if the photon propagator is replaced with the gluon propagator in the following manner:

$$iD_{F\mu\nu} \to iG_{\mu\nu} = i\lambda^4 g_{\mu\nu} \mathrm{P} \frac{1}{k^6} \tag{B11}$$

in the Feynman gauge, with

$$\mathrm{P}\frac{1}{k^6} = \frac{1}{2}\frac{1}{(k^2+i\epsilon)^3} + \frac{1}{2}\frac{1}{(k^2-i\epsilon)^3}\,. \tag{B12}$$

In coordinate space the gluon propagator is

$$T(A^1_\mu(x)A^1_\nu(y)) = i\lambda^4 g_{\mu\nu}\int d^4k\left(\mathrm{P}\frac{1}{k^6}\right)e^{-ik\cdot(x-y)} \tag{B13}$$

$$= \frac{i}{64\pi}\lambda^4 g_{\mu\nu}(x-y)^2\theta((x-y)^2) \tag{B14}$$

in the Feynman gauge, which suggests a relationship between our model and that of Feynman, Kislinger, and Ravndal[4] as stated previously.

The discussions of unitarity, causality, and quark confinement given in the text apply to this model with only superficial changes. The light-cone divergences encountered in the dipole model are not so extreme here. For example, the overall degree of light-cone divergence for one-loop diagrams is 3-3p [where p is the number of internal gluon lines; cf. Eq. (61)] and thus the lowest-order quark self-energy (Fig. 2) is only logarithmically divergent,

$$\Sigma(q) = \frac{\lambda^4 g^2}{16\pi^2}\,\mathrm{P}\frac{1}{q^4}\left\{\not{q}\left[\ln\left(\frac{\Lambda^2}{m^2}\right) - \frac{2q^2m^2}{(q^2-m^2)^2} + \frac{3q^4m^2-q^6}{(q^2-m^2)^3}\ln\left(\frac{q^2}{m^2}\right)\right] + 2m\left[\frac{q^2(q^2+m^2)}{(q^2-m^2)^2} - \frac{2q^4m^2}{(q^2-m^2)^3}\ln\left(\frac{q^2}{m^2}\right)\right]\right\}, \tag{B15}$$

where q is the quark four-momentum, m the quark mass, Λ^2 is a regulator mass, and P signifies that all singularities are to be taken in principal value.

Finally, we would like to note again that the deep-inelastic structure functions scale in this model with leading nonscaling corrections of $O(1/q^6)$, where q is the virtual-photon four-momentum. These corrections come from the diagrams of Fig. 4.

---

*Work supported in part by the National Science Foundation.

[1] A. Casher, J. Kogut, and L. Susskind, Phys. Rev. Lett. 31, 792 (1973).

[2] K. Wilson, Phys. Rev. D 10, 2445 (1974).

[3] A. Chodos, R. L. Jaffe, K. Johnson, C. B. Thorn, and V. F. Weisskopf, Phys. Rev. D 9, 3471 (1974).

[4] R. Feynman, M. Kislinger, and F. Ravndal, Phys. Rev. D 3, 2706 (1971).

[5] P. A. M. Dirac, *Lectures on Quantum Mechanics* (Yeshiva Univ., New York, 1964).

[6] K. Schwarzschild, Göttinger Nachrichten 128, 132 (1903); H. Tetrode, Z. Phys. 10, 317 (1922); A. D. Fokker, *ibid.* 58, 386 (1929); Physica (The Hague) 9, 33 (1929); 12, 145 (1932); J. Wheeler and R. P. Feynman, Rev. Mod. Phys. 17, 157 (1945); 21, 425 (1949).

[7] J. D. Bjorken and S. D. Drell, *Relativistic Quantum Fields* (McGraw-Hill, New York, 1965), p. 382.

[8] E. Salpeter and H. Bethe, Phys. Rev. 84, 1232 (1951).

[9] R. J. Eden, P. V. Landshoff, D. I. Olive, and J. C. Polkinghorne, *The Analytic S-Matrix* (Cambridge Univ. Press, Cambridge, 1966), p. 34.

[10] Suggested by Dr. D. K. Sinclair.

[11] J. Schwinger, Phys. Rev. 128, 2425 (1962).

[12] After this work was completed Dr. K. Subbarao pointed out that Professor E. C. G. Sudarshan has considered the use of action-at-a-distance interactions to remedy unitarity problems in indefinite-metric theories: E. C. G. Sudarshan, Fields and Quanta 2, 175 (1972).

[13] S. Coleman, in *Subnuclear Phenomena*, edited by A. Zichichi (Academic, New York, 1970), Part A, p. 282.

[14] R. J. Eden, P. V. Landshoff, D. I. Olive, and J. C. Polkinghorne, *The Analytic S-Matrix* (Ref. 9), p. 112.

[15] R. Jackiw and G. Preparata, Phys. Rev. Lett. 22, 975 (1969); S. Adler and W. Tung, *ibid.* 22, 978 (1969); S. Blaha, Phys. Rev. D 3, 510 (1971).

[16] This may be proved using the representation of Eq. (54) with $q=0$ (only principal-value propagators) and $I=1$. If we let $\alpha_i \to -\alpha_i$ for $i=1,2,\ldots,p$ then we can show that $I=-I$ and thus $I=0$. $\tilde{N}$ is arbitrary except that it can be written as a sum of terms which are homogeneous in the Feynman parameters. This condition can always be satisfied in perturbation theory.

Appendix C. PseudoQuantization Paper

This refereed paper is S. Blaha, Phys. Rev. **D17**, 994 (1978). Reprinted with the kind permission of Physical Review D.

PHYSICAL REVIEW D VOLUME 17, NUMBER 4 15 FEBRUARY 1978

Embedding classical fields in quantum field theories

Stephen Blaha*
Physics Department, Syracuse University, Syracuse, New York 13210
(Received 2 August 1976; revised manuscript received 7 November 1977)

We describe a procedure for quantizing a classical field theory which is the field-theoretic analog of Sudarshan's method for embedding a classical-mechanical system in a quantum-mechanical system. The essence of the difference between our quantization procedure and Fock-space quantization lies in the choice of vacuum states. The key to our choice of vacuum is the procedure we outline for constructing Lagrangians which have gradient terms linear in the field variables from classical Lagrangians which have gradient terms which are quadratic in field variables. We apply this procedure to model electrodynamic field theories, Yang-Mills theories, and a vierbein model of gravity. In the case of electrodynamics models we find a formalism with a close similarity to the coherent-soft-photon-state formalism of QED. In addition, photons propagate to $t = +\infty$ via retarded propagators. We also show how to construct a quantum field for action-at-a-distance electrodynamics. In the Yang-Mills case we show that a previously suggested model for quark confinement necessarily has gluons with principal-value propagation which allows the model to be unitary despite the presence of higher-order-derivative field equations. In the vierbein-gravity model we show that our quantization procedure allows us to treat the classical and quantum parts of the metric field in a unified manner. We find a new perturbation scheme for quantum gravity as a result.

I. INTRODUCTION

The relation between classical and quantum systems has been a subject of continuing interest over the years: First, in the original development of quantum mechanics, second, in the study of the classical limit and infrared divergences of quantum-electrodynamic processes,[1,2] and third, in recent attempts to construct strong-interaction models of quark confinement which are for the most part either classical field theory models in search of quantization[3] or quantized gluon models wherein quark confinement is a consequence of infrared behavior.[4,5]

We will describe a new quantization procedure (called pseudoquantization) for field theory which is the analog of Sudarshan's method for embedding a classical-mechanical system in a quantum-mechanical system. It can be used with advantage to either embed a classical field theory in a quantum field theory in such a way as to maintain the classical character of the embedded fields (while studying the interaction between the classical and quantum sectors on essentially the same footing), or to quantize a class of field theories, members of which have been used as models for gravity and as models for the strong interaction with quark confinement.[7-9]

We shall begin (Sec. II) by pseudoquantizing a classical simple harmonic oscillator. This case is of particular importance because of the analogy between the mode amplitudes of a quantum field and the coordinates of a set of simple harmonic oscillators which we will take advantage of in later sections.

In Sec. III we describe the pseudoquantization procedure for field theory. We apply it to electrodynamic models and show that the propagation of photons to $t=+\infty$ is necessarily retarded in this formalism. Further, we display a close analogy between the present formalism and the coherent-soft-photon-state formalism[10] of QED.

In Sec. IV we apply the pseudoquantization procedure to a classical Yang-Mills field. The resulting field theory (with a slight but important modification) has been used as a model for the strong interactions with quark confinement.[7-9] We also apply the pseudoquantization procedure to a vierbein model of gravity and obtain a new perturbation theory for quantum gravity.

In Sec. V we show that principal-value propagators naturally arise in certains sectors of pseudoquantized theories thus verifying an *ad hoc* procedure devised to unitarize a model of quark confinement.[7-9] We also show how to construct a quantum version of action-at-a-distance electrodynamics.

We shall now briefly outline the procedure for embedding a classical-mechanical system in a quantum system.[6] Consider a classical Hamiltonian system with one degree of freedom, and commuting canonical variables, x_1 and p_1, which have the equations of motion

$$\dot{x}_1 = -i[x_1, \hat{H}], \tag{1}$$

$$\dot{p}_1 = -i[p_1, \hat{H}], \tag{2}$$

where defining

$$\hat{H} = -i\left(\frac{\partial H(x_1,p_1)}{\partial p_1}\frac{\partial}{\partial x_1} - \frac{\partial H(x_1,p_1)}{\partial x_1}\frac{\partial}{\partial p_1}\right) \tag{3}$$

allows us to write Hamilton's equations in commutator form. With Sudarshan[6] we define

$$x_2 = i\frac{\partial}{\partial p_1} \tag{4}$$

and

$$p_2 = -i\frac{\partial}{\partial x_1} \tag{5}$$

so that

$$[x_1, x_2] = [p_1, p_2] = 0\,, \tag{6}$$

$$[x_1, p_2] = [x_2, p_1] = i\,, \tag{7}$$

and $\hat{H}$ can now be taken to be the operator

$$\hat{H} = \frac{\partial H(x_1, p_1)}{\partial p_1} p_2 + \frac{\partial H(x_1, p_1)}{\partial x_1} x_2\,. \tag{8}$$

It is now apparent that we can take the above quantities and equations of motion to describe a quantum mechanical system with two degrees of freedom in the "coordinate" representation where the "coordinates" are (x_1, p_1) and the canonical momenta are $\Pi = (p_2, -x_2)$. As we will see below the linearity of $\hat{H}$ in the momenta is crucial for the maintenance of the classical character of x_1 and p_1, and for the observability of the phase-space trajectory. Since we choose to identify the physical observables with the commutative algebra of the coordinate operators, x_1 and p_1, we are led to impose the superselection condition that the momenta, Π, are unobservable. As a result the Hamiltonian and other generators of canonical transformations, which are all linear in the momenta, are also unobservable. However, in each case there is an associated dynamical quantity which is observable.

The required unobservability of the momenta restricts the form of the interaction between a classical-made-quantum system and an inherently quantum system to

$$H_{\text{int}} = \Phi_1 x_2 + \Phi_2 p_2 + X\,, \tag{9}$$

where Φ_1, Φ_2, and X are functions of x_1, p_1, and the quantum system variables. The commutation relations of these functions are also constrained[6] by the superselection rule and the commutativity of the classical variables, x_1 and p_1, and their time derivatives. In the next section we will study the simple harmonic oscillator in order to exemplify the quantum-mechanical case described above and also for direct use in the field-theoretic generalizations of subsequent sections.

II. SIMPLE HARMONIC OSCILLATOR

In this section we discuss the embedding of a classical simple harmonic oscillator in a quantum system. We shall see that the space of states for the indefinite-metric classical-made-quantum system is far larger than the set of states of a classical harmonic oscillator. However, there is a subset of coherent states which may be placed in one-to-one correspondence with the classical harmonic-oscillator states. The classical-made-quantum oscillator is necessarily an indefinite-metric quantum theory for the simple physical reason that the classical bound states cannot have quantized energy levels. Indefinite-metric quantum theories normally have severe problems of physical interpretation. The present work raises the possibility of a partial resolution of some of these problems through a reinterpretation of an indefinite-metric quantum system as a system composed of a classical subsystem interacting with an essentially quantum subsystem of positive metric.

The classical simple harmonic oscillator of frequency ω has the Hamiltonian

$$\mathcal{H} = \frac{1}{2m}(p_1{}^2 + m^2\omega^2 x_1{}^2)\,, \tag{10}$$

and the motion is described by

$$x_1 = A\sin(\pi t + \delta)\,, \tag{11}$$

where A and δ are constants. To embed this classical system in a quantum-mechanical system we introduce the variables x_2 and p_2, and, using Eq. (8), obtain the quantum Hamiltonian

$$\hat{H} = \frac{1}{m} p_1 p_2 + m\omega^2 x_1 x_2\,. \tag{12}$$

We eliminate constants by defining (for $i = 1, 2$)

$$x_1 = \left(\frac{1}{m\omega}\right)^{1/2} Q_1\,, \tag{13}$$

$$p_1 = (m\omega)^{1/2} P_1\,, \tag{14}$$

and

$$\hat{H} = H\omega \tag{15}$$

so that

$$H = P_1 P_2 + Q_1 Q_2\,. \tag{16}$$

The raising and lowering operators are defined by

$$a_j = \frac{1}{\sqrt{2}}(Q_j + iP_j)\,, \tag{17}$$

and

$$a_j^\dagger = \frac{1}{\sqrt{2}}(Q_j - iP_j) \tag{18}$$

for $j = 1, 2$. They have the commutation relations

$$[a_i, a_j] = [a_i^\dagger, a_j^\dagger] = 0\,, \tag{19}$$

$$[a_i, a_j^\dagger] = 1 - \delta_{ij} \tag{20}$$

for $i,j=1,2$. As a result H is seen to have the form

$$H=\tfrac{1}{2}(a_1a_2^\dagger+a_2a_1^\dagger+a_1^\dagger a_2+a_2^\dagger a_1)\,. \tag{21}$$

The number operators are defined by

$$N_1=a_2a_1^\dagger \tag{22}$$

and

$$N_2=a_2^\dagger a_1 \tag{23}$$

and are not Hermitian. However, their sum is Hermitian and we see that

$$H=N_1+N_2\,. \tag{24}$$

The number operators have the following commutation relations with the raising and lowering operators:

$$N_ia_j=a_j(N_i+\delta_{ij}-1) \tag{25}$$

and

$$N_ia_j^\dagger=a_j^\dagger(N_i-\delta_{ij}+1) \tag{26}$$

for $i,j=1,2$.

Up to this point we have maintained a symmetry of the dynamics under the exchange of the subscripts, $1 \leftrightarrow 2$. Now we must break that symmetry by choosing a vacuum state which is an eigenstate of Q_1 and P_1 or alternately a_1 and $a_1^\dagger$. The commutativity of Q_1 and P_1 permit this. The observability of Q_1 and P_1 for all time requires it. So we define

$$a_1^\dagger|0\rangle=a_1|0\rangle=0\,. \tag{27}$$

As a result $a_2|0\rangle\neq 0$ and $a_2^\dagger|0\rangle\neq 0$. The eigenstates of the number operators are

$$|n_+,n_-\rangle=(a_2^\dagger)^{n_+}(a_2)^{n_-}|0,0\rangle \tag{28}$$

and satisfy

$$N_1|n_+,n_-\rangle=-n_-|n_+,n_-\rangle\,, \tag{29}$$

$$N_2|n_+,n_-\rangle=n_+|n_+,n_-\rangle\,, \tag{30}$$

so that

$$H|n_+,n_-\rangle=(n_+-n_-)|n_+,n_-\rangle\,. \tag{31}$$

The lack of a lower bound to the energy spectrum is in a sense a problem but a necessary one in that it leads to the possibility of bound states with a continuous energy spectrum—a requirement of a faithful representation of the classical oscillator states. There is a subset of coherent states which can be put in a one-to-one relation with the set of classical oscillator states. The defining property of that subset is that its elements are eigenstates of the operators a_1 and $a_1^\dagger$. If we expand an element of that subset in terms of the number eigenstates

$$|z\rangle=\sum_{n_+,n_-=0}^{\infty} f(z\,|n_+,n_-)|n_+,n_-\rangle \tag{32}$$

and use

$$a_1^\dagger|n_+,n_-\rangle=-n_-|n_+,n_--1\rangle\,, \tag{33}$$

$$a_1|n_+,n_-\rangle=n_+|n_+-1,n_-\rangle \tag{34}$$

to evaluate the eigenvalue equations

$$a_1|z\rangle=iz^*|z\rangle\,, \tag{35}$$

$$a_1^\dagger|z\rangle=-iz|z\rangle\,, \tag{36}$$

we find

$$f(z\,|n_+,n_-)=\frac{C(iz^*)^{n_+}(iz)^{n_-}}{n_+!\,n_-!}\,, \tag{37}$$

where C is a constant. As a result

$$|z\rangle=C\exp[i(za_2+z^*a_2^\dagger)]|0,0\rangle\,. \tag{38}$$

We shall call the $|z\rangle$ states coherent states because of their close formal resemblance to the coherent states used in the study of the classical limit of harmonic oscillators, and of quantum electrodynamics[11] (which were eigenstates of the lowering operator but not of the raising operator).

Since $[H,a_1]=-a_1$, and $[H,a_1^\dagger]=a_1^\dagger$, it is clear that the (x_1,p_1) phase-space trajectory is sharp on the set of coherent $|z\rangle$ states. The classical trajectory represented by the state $|z\rangle$ is easily seen to be

$$x_1=\left(\frac{2}{m\omega}\right)^{1/2}R\sin(\omega t+\delta) \tag{39}$$

and

$$p_1=(2m\omega)^{1/2}R\cos(\omega t+\delta)\,, \tag{40}$$

where $z=Re^{i\delta}$. The linearity of H in the "momenta", $\Pi=(p_2,-x_2)$, is crucial for the observability of the phase-space trajectory. In fact, the linearity of all generators of canonical transformations in the momenta is necessary if the canonical transformations are not to take states out of the subset of coherent states.

The superselection rule which follows from the unobservability of the momenta, Π, is best approached by a consideration of the momentum-and coordinate-space representations of the coherent states. In the coordinate-space representation we find that Eqs. (35) and (36) give

$$\left[\left(\frac{m\omega}{2}\right)^{1/2}x_1+i\left(\frac{1}{2m\omega}\right)^{1/2}p_1\right]\langle x_1p_1|z\rangle=iz^*\langle x_1p_1|z\rangle \tag{41}$$

and

$$\left[\left(\frac{m\omega}{2}\right)^{1/2}x_1-i\left(\frac{1}{2m\omega}\right)^{1/2}p_1\right]\langle x_1p_1|z\rangle=-iz\langle x_1p_1|z\rangle\,, \tag{42}$$

so that

$$\langle x_1 p_1 | z\rangle = \sqrt{2}\,\delta\left(x_1 - \left(\frac{2}{m\omega}\right)^{1/2} \mathrm{Im} z\right) \times \delta(p_1 - (2m\omega)^{1/2}\mathrm{Re} z). \quad (43)$$

We have normalized $\langle x_1 p_1|z\rangle$ so that

$$\langle z'|z\rangle = \int_{-\infty}^{\infty} dx_1 dp_1 \langle z'|x_1 p_1\rangle\langle x_1 p_1|z\rangle = \delta(\mathrm{Re} z - \mathrm{Re} z')\delta(\mathrm{Im} z - \mathrm{Im} z'). \quad (44)$$

In momentum space Eqs. (35) and (36) lead to the differential equations

$$\left[\left(\frac{m\omega}{2}\right)^{1/2} i\frac{d}{dp_2} + \left(\frac{1}{2m\omega}\right)^{1/2}\frac{d}{dx_2}\right]\langle x_2 p_2|z\rangle = iz^*\langle x_2 p_2|z\rangle \quad (45)$$

and

$$\left[\left(\frac{m\omega}{2}\right)^{1/2} i\frac{d}{dp_2} - \left(\frac{1}{2m\omega}\right)^{1/2}\frac{d}{dx_2}\right]\langle x_2 p_2|z\rangle = -iz\langle x_2 p_2|z\rangle. \quad (46)$$

They are easily integrated to give

$$\langle x_2 p_2|z\rangle = \frac{1}{\sqrt{2\pi}}\exp\left[-ip_2\left(\frac{2}{m\omega}\right)^{1/2}\mathrm{Im} z + ix_2(2m\omega)^{1/2}\mathrm{Re} z\right] \quad (47)$$

with the normalization condition

$$\langle z'|z\rangle = \int_{-\infty}^{\infty} dx_2 dp_2 \langle z'|x_2 p_2\rangle\langle x_2 p_2|z\rangle = \delta(\mathrm{Re} z - \mathrm{Re} z')\delta(\mathrm{Im} z - \mathrm{Im} z'). \quad (48)$$

The transformation function between the two representations is

$$\langle x_1 p_1|x_2 p_2\rangle = \frac{1}{2\pi}\exp(+ip_2 x_1 - ip_1 x_2), \quad (49)$$

so that

$$\langle x_1 p_1|z\rangle = \int_{-\infty}^{\infty} dx_2 dp_2 \langle x_1 p_1|x_2 p_2\rangle\langle x_2 p_2|z\rangle. \quad (50)$$

Each coherent state, $|z\rangle$, is a superselection sector in itself. There is no measurable dynamical variable $F = F(a_1, a_1^\dagger)$ which connects different states:

$$\langle z'|F(a_1, a_1^\dagger)|z\rangle = F(iz^*, -iz)\delta^2(z - z'). \quad (51)$$

This reflects the lack of a superposition principle in classical mechanics.

The operator formalism for coherent states is incomplete in that we have not defined an inner product. To remedy this deficiency we define the vacuum dual to $|0,0\rangle$ to satisfy

$$\langle 0,0|a_2 = \langle 0,0|a_2^\dagger = 0 \quad (52)$$

with $\langle 0,0|0,0\rangle = 1$. The dual state corresponding to the physical state, z, we define to be

$$\langle z| = \langle 0,0|\delta(ia_1 + z^*)\delta(ia_1^\dagger - z) \equiv \langle 0,0|\int_{-\infty}^{\infty}\frac{d\alpha\, d\beta}{(2\pi)^2}\exp[i\alpha(\mathrm{Im} z - 2^{-1/2}Q_1) + i\beta(\mathrm{Re} z - 2^{-1/2}P_1)] \quad (53)$$

so that Eqs. (48) and (51) follow if we choose $C = 1$.

Sometimes the dynamical state of a classical system is incompletely known and one only has a set of probabilities that the system is at a particular phase-space point at $t = 0$. If we let $P(z)$ be the probability that the system is at a phase-space point corresponding to z (as defined above), then using the properties

$$P(z) \geq 0, \quad \int d^2z\, P(z) = 1 \quad (54)$$

one sees that a density operator

$$\rho\delta^2(0) = \int d^2z\, |z\rangle P(z)\langle z| \quad (55)$$

may be defined which satisfies

$$\mathrm{Tr}\rho = 1 \quad (56)$$

and

$$\langle z'|\rho|z'\rangle \equiv \lim_{z''\to z'}\langle z''|\rho|z'\rangle = P(z'). \quad (57)$$

The mean value of an observable $A = A(a_1, a_1^\dagger)$ is given by

$$\langle A\rangle = \mathrm{Tr}\rho A = \int d^2z\, A(iz^*, -iz)P(z), \quad (58)$$

and one can develop a formalism similar to the density-matrix formalism of quantum mechanics.

We now turn to a closer investigation of the relation of the pseudoquantum mechanics discussed above and true quantum-mechanical systems. We shall be particularly interested in the relation of the coherent states described above and the coherent states of a quantum-mechanical harmonic oscillator—to which they bear such a remarkable resemblance. We shall see that the pseudoquantum oscillator system is equivalent to an indefinite-metric quantum system composed of a harmonic oscillator (thus the connection to the coherent-state quantum oscillator formalism) and an "inverted" oscillator to be described below.

Let us define the following rotated raising and lowering operators in terms of the operators defined in Eqs. (17) and (18):

$$b_1 = a_1\cos\theta + a_2\sin\theta, \quad (59)$$

$$b_2 = -a_1\sin\theta + a_2\cos\theta. \quad (60)$$

Their commutation relations are

$$[b_1, b_1^\dagger] = \sin(2\theta), \tag{61}$$

$$[b_2, b_2^\dagger] = -\sin(2\theta), \tag{62}$$

$$[b_2, b_1^\dagger] = [b_1, b_2^\dagger] = \cos(2\theta) \tag{63}$$

with all other commutators equal to zero. The Hamiltonian of Eq. (21) becomes

$$H = \tfrac{1}{2}(\{b_1, b_1^\dagger\} - \{b_2, b_2^\dagger\}) \sin(2\theta) + \tfrac{1}{2}(\{a_1, a_2^\dagger\} + \{a_2, a_1^\dagger\}) \cos(2\theta), \tag{64}$$

where $\{u, v\} = uv + vu$.

Now θ is an arbitrary angle and it is obvious that choosing $\theta = 0$ gives the commutation relations and Hamiltonian studied above. However, the choice $\theta = \pi/4$ results in a new form for H and the commutation relations, which can be interpreted as a harmonic oscillator (the b_1 and $b_1^\dagger$ sector) and an "inverted" harmonic oscillator (the b_2 and $b_2^\dagger$ sector) where the commutator and b_2 terms in the Hamiltonian have the wrong sign. The commutativity of the oscillator raising and lowering operators with the inverted oscillator raising and lowering operators leads to a simple factorization of the coherent states which lays bare the basic of the close similarity of form for our coherent states and the coherent states of a quantum oscillator[10]:

$$|z\rangle = \frac{1}{\sqrt{2\pi}} \exp\left[\frac{i}{\sqrt{2}}(z b_1 + z^* b_1^\dagger)\right] \times \exp\left[\frac{i}{\sqrt{2}}(z b_2 + z^* b_2^\dagger)\right] |0, 0\rangle, \tag{65}$$

while the coherent state of Ref. 11 has the form

$$|\alpha\rangle = \exp(\alpha b^\dagger - \alpha^* b) |0\rangle, \tag{66}$$

where α is a complex numer and $[b, b^\dagger] = 1$. It should be remembered that our choice of vacuum state such that $a_1|0,0\rangle = a_1^\dagger|0,0\rangle = 0$ obviates a simple direct relationship.

Since we have uncovered an interesting relation between a classical-made-quantum system and a "quantum" system of indefinite metric the possibility of reinterpreting indefinite-metric quantum systems as systems containing classical subsystems naturally arises.

III. EMBEDDING OF CLASSICAL FIELDS

In this section we shall discuss the embedding of a classical field theory in a quantum field theory. We shall study the embedding in detail for a scalar field and then describe the features of a classical-made-quantum electrodynamics which we shall call pseudoquantum electrodynamics for the sake of brevity.

Consider a classical field, $\phi_1(x)$, with canonically conjugate momentum, $\pi_1(x)$, and Hamiltonian equations of motion

$$\frac{d}{dt}\phi_1(x) = \frac{\delta \hat{H}}{\delta \pi_1(x)}, \tag{67}$$

$$\frac{d}{dt}\pi_1(x) = \frac{-\delta H}{\delta \phi_1(x)}, \tag{68}$$

where $\hat{H}$ is the Hamiltonian. We wish to define a "quantum" Hamiltonian, H, which allows us to rewrite Eqs. (67) and (68) in commutator form:

$$\frac{d}{dt}\phi_1(x) = i[H, \phi_1(x)], \tag{69}$$

$$\frac{d}{dt}\pi_1(x) = i[H, \pi_1(x)]. \tag{70}$$

Equations (69) and (70) are satisfied if

$$H = \int d^3x \left[\frac{\delta H}{\delta \pi_1(x)} \frac{1}{i} \frac{\delta}{\delta \phi_1(x)} - \frac{\delta H}{\delta \phi_1(x)} \frac{1}{i} \frac{\delta}{\delta \pi_1(x)}\right]. \tag{71}$$

We now formally define

$$\phi_2(x) = i\frac{\delta}{\delta \pi_1(x)} \tag{72}$$

and

$$\pi_2(x) = -i\frac{\delta}{\delta \phi_1(x)}, \tag{73}$$

so that

$$H = \int d^3x \left[\frac{\delta \hat{H}}{\delta \pi_1(x)} \pi_2(x) + \frac{\delta \hat{H}}{\delta \phi_1(x)} \phi_2(x)\right]. \tag{74}$$

The fields satisfy the equal-time commutation relations

$$[\phi_i(x), \pi_j(y)] = i(1 - \delta_{ij})\delta^3(\vec{x} - \vec{y}), \tag{75}$$

$$[\phi_i(x), \phi_j(y)] = 0, \tag{76}$$

$$[\pi_i(x), \pi_j(y)] = 0, \tag{77}$$

where δ_{ij} is the Kronecker δ.

We note that the linearity of H in ϕ_2 and π_2 is necessary to maintain the classical character of ϕ_1 and π_1. This is best seen by an examination of Eqs. (69) and (70) and the corresponding Hamiltonian equations for ϕ_2 and π_2. (Other generators of canonical transformations are also linear in π_2 and ϕ_2.)

$\phi_2(x)$ and $\pi_2(x)$ will not be observables on the set of physical states, so that $\phi_1(x)$ and $\pi_1(x)$ will both be sharp on the set of physical states and satisfy superselection rules.

If we wish to couple the classical field to a truly quantum system and maintain the classical nature of the field then certain restrictions exist on the form of the total Hamiltonian H_{tot} and on the commutation relations of the various terms occurring in it. First, the coupling must satisfy the requirement that H_{tot} is linear in $\phi_2(x)$ and $\pi_2(x)$. If we denote the quantum fields by ψ and write the general form of the Hamiltonian as

$$H_{tot}=H+H_Q(\psi)+H_{int}\,, \tag{78}$$

where H is given by Eq. (74), $H_Q(\psi)$ depends only on the quantum fields, ψ, and

$$H_{int}=\int d^3x[\tilde{A}(\phi_1,\pi_1,\psi)\phi_2(x) + \tilde{B}(\phi_1,\pi_1,\psi)\pi_2(x) + \tilde{C}(\phi_1,\pi_1,\psi)]\,, \tag{79}$$

then we can rearrange the Hamiltonian so that

$$H_{tot}=\int d^3x[A(\phi_1,\pi_1,\psi)\phi_2(x) + B(\phi_1,\pi_1,\psi)\pi_2(x) + C(\phi_1,\pi_1,\psi)]\,, \tag{80}$$

where

$$A=\frac{\delta\hat{H}}{\delta\phi_1(x)}+\tilde{A}\,, \tag{81}$$

$$B=\frac{\delta\hat{H}}{\delta\pi_1(x)}+\tilde{B}\,, \tag{82}$$

and

$$C=\tilde{C}+\mathcal{H}_Q \tag{83}$$

with $H_Q=\int d^3x\,\mathcal{H}_Q$. An examination of the equations of motion of $\phi_1(x)$, $\pi_1,(x)$, and ψ,

$$\frac{d}{dt}\phi_1=B(\phi_1,\pi_1,\psi)\,, \tag{84}$$

$$\frac{d}{dt}\pi_1=A(\phi_1,\pi_1,\psi)\,, \tag{85}$$

$$\frac{d}{dt}\psi=i[H_{tot},\psi]\,, \tag{86}$$

and the second time derivatives of ϕ_1 and π_1, such as

$$\frac{d^2}{dt^2}\phi_1(x)=i[H,B] = \int d^3y\left(-A\frac{\delta B}{\delta\pi_1(y)}+B\frac{\delta B}{\delta\phi_1(y)}+i\phi_2(y)[A,B] + i\pi_2(y)[B(y),B(x)]+i[C,B]\right), \tag{87}$$

leads us to require the equal-time commutation relations

$$[A(x),A(y)]=[A(x),B(y)]=[B(x),B(y)]=0\,, \tag{88}$$

where $A(x)=A(\phi_1(x),\pi_1(x),\psi(x))$, etc., so that $\phi_1(x)$ and $\pi_1(x)$ are independent of ϕ_2 and π_2 and hence observable for all time. An examination of higher time derivatives of ϕ_1 and π_1 lead to further restrictions on the equal-time commutation relations of A, B, and C. Examples are

$$[A,[C,B]]=0\,, \tag{89}$$

$$[B,[C,B]]=0\,, \tag{90}$$

$$[A,[C,[C,[C,B]]]]=0\,, \tag{91}$$

etc. A sufficient condition for satisfying all relations of this class consists of having equal-time commutation relations with the form

$$[A,C]=F_1(A,B,\phi_1,\pi_1) \tag{92}$$

and

$$[B,C]=F_2(A,B,\phi_1,\pi_1)\,. \tag{93}$$

Finally, we note that another obvious requirement [cf. Eqs. (84) and (85)] for the observability of ϕ_1 and π_1 is that A and B depend only on an (equal-time) commutative subset of the quantum field variables, ψ.

The above restrictions on the equal-time commutation relations have a direct interpretation in terms of Feynman diagrams for quantum corrections to the classical field behavior. For example, consider the interaction of the classical field sector with a scalar quantum field, ψ, expressed in the interaction

$$H_{int}=g\phi_2(x)\psi^2(x). \tag{94}$$

If $H_Q(\psi)$ is the conventional free Klein-Gordon Hamiltonian, then we find that Eq. (92) is not satisfied so that the Green's function for the classical ϕ_1 field receives quantum corrections from vacuum polarization loops of ψ particles and thus loses its classical character.

We now define a Lagrangian appropriate to our pseudoquantum field theory and then verify the reasonableness of our definition, and the pseudoquantization procedure described above, by studying the equivalent path-integral formulation. The Lagrangian corresponding to the pseudoquantum Hamiltonian, H, is

$$L=\int d^3x(\pi_1\mathring{\phi}_2+\pi_2\mathring{\phi}_1)-H\,, \tag{95}$$

where $L=L(\phi_1,\mathring{\phi}_1,\phi_2,\mathring{\phi}_2)$ and

$$\pi_1=\frac{\delta L}{\delta\mathring{\phi}_2}\,, \tag{96}$$

$$\pi_2 = \frac{\delta L}{\delta \dot{\phi}_1} . \tag{97}$$

The vacuum-vacuum transition amplitude for the field theory corresponding to the H_{tot} of Eq. (78) will be shown to be

$$W = \int \prod_x d\phi_1(x) d\phi_2(x) d\pi_1(x) d\pi_2(x) d\psi(x) \exp(iS) , \tag{98}$$

where $S = \int dt\, L_{tot}$ up to external source terms. We begin by considering the vacuum-vacuum transition amplitude corresponding to H_Q,

$$W_Q = \int \prod_x d\psi(x) \exp(iS_Q) , \tag{99}$$

where ϕ_1 has the character of an external source.

We can now introduce the classical behavior of the ϕ_1 field through functional δ functions

$$\int \prod_x d\psi(x) d\phi_1(x) d\pi_1(x) \delta(B(\phi_1, \pi_1, \psi) - \dot{\phi}_1) \times \delta(A(\phi_1, \pi_1, \psi) + \dot{\pi}_1) e^{iS_Q} , \tag{100}$$

which can be put in the form

$$\int \prod_x d\phi_1(x) d\pi_1(x) d\phi_2(x) d\pi_2(x) \times \exp\left\{ i \int d^4x [(\dot{\phi}_1 - B)\pi_2 - (\dot{\pi}_1 + A)\phi_2] + iS_Q \right\} . \tag{101}$$

After performing a partial integration on the $\dot{\pi}_1\phi_2$ term and discarding a surface term we see that the definition of L in Eq. (95) is correct and that the vacuum-vacuum transition amplitude is indeed given by Eq. (98).

The restrictions on the commutation relations of the various terms in the H_{tot} [expressed in Eqs. (88)–(93)] translate into the requirement that the "quantum completion"[11] of the ϕ_2 field does not take place, i.e., that all N-point functions of the ϕ_2 field are zero:

$$\frac{\delta^n W}{\delta J_2(x_1) \delta J_2(x_2) \cdots \delta J_2(x_n)} = 0 , \tag{102}$$

where J_2 is an external source coupled to ϕ_2.

We now discuss the embedding of a free classical Klein-Gordon field in a quantum field theory. The Lagrangian density is

$$\mathcal{L} = \frac{\partial \phi_1}{\partial x^\mu} \frac{\partial \phi_2}{\partial x_\mu} - m^2 \phi_1 \phi_2 , \tag{103}$$

from which one obtains the Euler-Lagrange equations (for $i = 1, 2$)

$$(\Box + m^2)\phi_i(x) = 0. \tag{104}$$

The canonical momenta are (note that π_2 is conjugate to ϕ_1, etc.)

$$\Pi_i = \dot{\phi}_i \tag{105}$$

for $i = 1, 2$ with the equal-time commutation relations given by Eqs. (75)–(77). We expand the fields in Fourier integrals:

$$\phi_1(\vec{x}, t) = \int d^3k [a_1(k) f_k(x) + a_1^\dagger f_k^*(x)] \tag{106}$$

and

$$\phi_2(\vec{x}, t) = \int d^3k [a_2(k) f_k(x) + a_2^\dagger(k) f_k^*(x)] , \tag{107}$$

where

$$f_k(x) = (2\pi)^{-3/2} (2\omega_k)^{-1/2} e^{-ik\cdot x} \tag{108}$$

with $\omega_k = (\vec{k}^2 + m^2)^{1/2}$. The Fourier component operators satisfy the commutation relations

$$[a_i(k), a_j^\dagger(k')] = (1 - \delta_{ij}) \delta^3(\vec{k} - \vec{k}') \tag{109}$$

and

$$[a_i(k), a_j(k')] = [a_i^\dagger(k), a_j^\dagger(k')] = 0 \tag{110}$$

for $i, j = 1, 2$.

In terms of the Fourier coefficients

$$H \equiv \int d^3x (\dot{\phi}_1 \dot{\phi}_2 + \vec{\nabla}\phi_1 \cdot \vec{\nabla}\phi_2 + m^2 \phi_1 \phi_2) \tag{111}$$

becomes

$$H = \int d^3k\, \omega_k [\{a_1(k), a_2^\dagger(k)\} + \{a_2(k), a_1^\dagger(k)\}] . \tag{112}$$

The analogy between the mode amplitudes of the fields and the raising and lowering operators of the simple harmonic oscillator has been previously remarked. We can therefore use the considerations of Sec. II to establish the spectrum of physical states. The defining properties of a physical state are that $\phi_1(x)$ and $\pi_1(x)$ are sharp on it for all time:

$$\phi_1(x) |\Phi, \Pi\rangle = \Phi(x) |\Phi, \Pi\rangle \tag{113}$$

and

$$\pi_1(x) |\Phi, \Pi\rangle = \Pi(x) |\Phi, \Pi\rangle , \tag{114}$$

where $\Phi(x)$ and $\Pi(x)$ are c-number functions of x:

$$\Phi(x) = \int d^3k [\alpha(k) f_k(x) + \alpha^*(k) f_k^*(x)] \tag{115}$$

and

$$\Pi(x) = -i \int d^3k\, \omega_k [\alpha(k) f_k(x) - \alpha^*(k) f_k^*(x)] \tag{116}$$

with $\alpha(k)$ a c-number function of k.

As a result we are led to define a set of physical states, $|\alpha\rangle$, which are in one-to-one correspondence with the classical solutions of the Klein-Gordon equation and satisfy

$$a_1(k)|\alpha\rangle = \alpha(k)|\alpha\rangle, \tag{117}$$

$$a_1^\dagger(k)|\alpha\rangle = \alpha^*(k)|\alpha\rangle. \tag{118}$$

In analogy with the states of the simple harmonic oscillator (Sec. II) we further define

$$|\alpha\rangle = C\exp\left\{\int d^3k'[\alpha(k')a_2^\dagger(k') - \alpha^*(k')a_2(k')]\right\}|0\rangle, \tag{119}$$

where the vacuum state, $|0\rangle$, satisfies

$$a_1(k)|0\rangle = a_1^\dagger(k)|0\rangle = 0. \tag{120}$$

The physical states, $|\alpha\rangle$, lie in a space which is the infinite tensor product of single-mode spaces. While ϕ_1 and π_1 are sharp for all time on the subset of physical states, we see that ϕ_2 and π_2 are not and, in fact, when applied to a physical state map it into an unphysical state. The superselection rules are embodied in

$$\langle\alpha'|\mathcal{O}|\alpha\rangle = \mathcal{O}_\alpha\delta^2(\alpha-\alpha'), \tag{121}$$

where $\mathcal{O}$ is the operator corresponding to any observable, $\mathcal{O}_\alpha$ is its eigenvalue for the state $|\alpha\rangle$, and $\delta^2(\alpha-\alpha')$ is a functional δ function in the real and imaginary parts of $\alpha-\alpha'$. The functional δ functions have their origin in the definition of the dual set of physical states. We define the dual vacuum state $\langle 0|$ by

$$\langle 0|a_2(k) = 0 \tag{122a}$$

and

$$\langle 0|a_2^\dagger(k) = 0 \tag{122b}$$

for all k with $\langle 0|0\rangle = 1$. The dual state corresponding to $\alpha(k)$ we define by

$$\langle\alpha| = \langle 0|\prod_k \delta(\alpha(k) - a_1(k))\delta(\alpha^*(k) - a_1^\dagger(k)) \equiv \langle 0|\delta(\alpha - a_1)\delta(\alpha^* - a_1^\dagger), \tag{123}$$

so that

$$\langle\alpha'|\alpha\rangle = \delta^2(\alpha'-\alpha) \tag{124}$$

if $C=1$.

We have now established a procedure for embedding a classical field in a quantum field theory. Given a Lagrangian, L, for a classical field theory describing a field $\phi_1(x)$, the Lagrangian density for the pseudoquantum field theory, $\mathcal{L}_{PQ}$ is

$$\mathcal{L}_{PQ}(\phi_1,\dot\phi_1,\phi_2,\dot\phi_2) = \frac{\delta L}{\delta\phi_1(x)}\phi_2(x) + \frac{\delta L}{\delta\dot\phi_1(x)}\pi_2(x) \tag{125}$$

up to a divergence with

$$\pi_2(x) = \frac{\delta}{\delta\dot\phi_1(x)}\int d^3x\,\mathcal{L}_{PQ}. \tag{126}$$

In the case of a classical electromagnetic field interacting with a quantum electron field, one pseudoquantum model, which describes some electromagnetic processes, has the Lagrangian

$$\mathcal{L} = -\tfrac{1}{2}F^1_{\mu\nu}F^2_{\mu\nu} + \bar\psi(i\nabla\!\!\!\!/ - eA\!\!\!/_1 - m_0)\psi, \tag{127}$$

where $A^1_\mu(x)$ is the classical electromagnetic field, ψ is the electron field, $A^2_\mu(x)$ is the unobservable auxiliary field, and $F^i_{\mu\nu} = \partial_\nu A^i_\mu - \partial_\nu A^i_\mu$ for $i=1,2$. Although our interpretation of the free electromagnetic part of the Lagrangian, $-\frac{1}{2}F^1_{\mu\nu}F^2_{\mu\nu}$, is new, the actual form of this term appeared some time ago in a generalization of electrodynamics by Mie,[12] and was recently used in an Abelian prototype model for quark confinement.[8] The equations of motion are

$$\partial^\mu F^1_{\mu\nu} = 0, \tag{128}$$

$$\partial^\mu F^2_{\mu\nu} + eJ_\nu = 0, \tag{129}$$

and

$$(i\nabla\!\!\!\!/ - eA\!\!\!/^1 - m)\psi = 0. \tag{130}$$

The canonical momentum which is conjugate to A^1_μ is

$$\Pi^2_\mu = F^2_{0\mu} \tag{131}$$

and that conjugate to A^2_μ is

$$\Pi^1_\mu = F^1_{0\mu}. \tag{132}$$

We take A^1_μ and Π^1_μ to be classical fields which are observable for all time. A^2_μ and Π^2_μ are not observable. Note that $\mathcal{L}$ is invariant under the independent gauge transformations

$$A^1_\mu \to A^1_\mu + \partial_\mu\Lambda^1(x) \tag{133}$$

and

$$A^2_\mu \to A^2_\mu + \partial_\mu\Lambda^2(x). \tag{134}$$

Since $\Pi^1_0 = \Pi^2_0 = 0$, it is apparent that A^1_0 and A^2_0 are c numbers. If we chose the Coulomb gauge for A^1_μ,

$$\vec\nabla\cdot\vec{A}^1 = 0, \tag{135}$$

and for A^2_μ,

$$\vec\nabla\cdot\vec{A}^2 = 0, \tag{136}$$

then we can establish the equal-time commutation relations

$$[\Pi_i^a(\vec{x},t),A_j^b(\vec{y},t)]=i(1-\delta_{ab})\times\int\frac{d^3k}{(2\pi)^3}e^{i\vec{k}\cdot(\vec{x}-\vec{y})}\left(\delta_{ij}-\frac{k_ik_j}{|\vec{k}|^2}\right)=i(1-\delta_{ab})\delta_{ij}^{tr}(\vec{x}-\vec{y}) \tag{137}$$

for $a,b=1,2$ and $i,j=1,2,3$.

This pseudoquantum field theory describes the dynamics of quantum electron fields interacting with a free, classical electromagnetic field. A typical perturbation theory matrix element would have the form

$$\langle\mathcal{A}',0|T(\bar{\psi}(x)J^{\mu_1}(x_1)A^1_{\mu_1}(x_1)J^{\mu_2}(x_2)A^1_{\mu_2}(x_2)\cdots J^{\mu_n}(x_n)A^1_{\mu_n}(x_n)\psi(y))|\mathcal{A},0\rangle, \tag{138}$$

where $|\mathcal{A},0\rangle$ is the tensor product of an electron vacuum state and an electromagnetic state corresponding to the classical field $\mathcal{A}_\mu(z)$. Because $A^1_\mu(x)$ is sharp on this state, the matrix element becomes

$$\langle 0|T(\bar{\psi}(x)J^{\mu_1}(x_1)\cdots J^{\mu_n}(x_n)\psi(y))|0\rangle\mathcal{A}_{\mu_1}(x_1)\mathcal{A}_{\mu_2}(x_2)\cdots\mathcal{A}_{\mu_n}(x_n) \tag{139}$$

modulo a functional δ function in $\mathcal{A}'-\mathcal{A}$. Thus this model is equivalent to a quantized electron field interacting with an external electromagnetic field.

Another possibility for a model electrodynamics is realized by letting the interaction term in Eq. (127) above be replaced with

$$L_{\text{int}}=-e\bar{\psi}A_2\psi. \tag{140}$$

Because the equivalent of the equal-time commutation relation, Eq. (92), is not true in this model, the A^1_μ field loses its purely classical character due to quantum corrections. However, this model may be of value for the study of the modification of the A^1_μ field resulting from the emission of many soft photons by a current.

Since vacuum polarization effects modify the electromagnetic field in this case we define in-field eigenstates (in the transverse gauge) by

$$\vec{A}^1_{\text{in}}|\mathcal{A}\rangle_{\text{in}}=\vec{\mathcal{A}}_{\text{in}}|\mathcal{A}\rangle_{\text{in}}, \tag{141}$$

where

$$|\mathcal{A}\rangle_{\text{in}}=\exp\left[\int d^3k\sum_{\lambda=1}^{2}(\alpha(k,\lambda)a_2^\dagger(k,\lambda)-\alpha^*(k,\lambda)a_2(k,\lambda))\right]|0\rangle \tag{142}$$

and

$$\vec{\mathcal{A}}_{\text{in}}=\int d^3k\sum_{\lambda=1}^{2}\vec{\epsilon}(k,\lambda)[\alpha(k,\lambda)f_k(x)+\alpha^*(k,\lambda)f_k^*(x)] \tag{143}$$

with

$$\vec{A}^i_{\text{in}}=\int d^3k\sum_{\lambda=1}^{2}\vec{\epsilon}(k,\lambda)[a_i(k,\lambda)f_k(x)+a_i^\dagger(k,\lambda)f_k^*(x)] \tag{144}$$

for $i=1,2$. The vacuum state is defined by

$$a_1(k,\lambda)|0\rangle=a_1^\dagger(k,\lambda)|0\rangle=0$$

for all k,λ. The interacting field, $\vec{A}^1$, is apparently not sharp on $|\mathcal{A}\rangle_{\text{in}}$ but is sharp on

$$|\mathcal{A}\rangle=U^{-1}(t,-\infty)|\mathcal{A}\rangle_{\text{in}}, \tag{145}$$

where

$$U(t,-\infty)=T\left(\exp\left[-i\int_{\infty}^{t}d^4x\,H_{\text{int}}(A^2_{\text{in}},\psi_{\text{in}})\right]\right) \tag{146}$$

because

$$\vec{A}^1(\vec{x},t)=U^{-1}(t,-\infty)\vec{A}^1_{\text{in}}(\vec{x},t)U(t,-\infty). \tag{147}$$

With these preliminaries completed, the study of physical processes within the framework of these models is now possible, although we shall not pursue it in this report.

Before turning to a discussion of non-Abelian gauge field theories, it is worth noting that the choice of vacuum state we have made necessitates a redefinition of normal-ordering. By normal-ordering a Lagrangian term we shall mean that the observable fields (to which we have consistently appended the superscript or subscript one) are to be placed to the right, and unobservable fields, labeled by two, are to be placed to the left. Thus Wick's theorem (with our definition of normal-ordering) becomes in the case of two fields

$$\begin{aligned}T(\phi_{1\,\text{in}}(x_1)\phi_{2\,\text{in}}(x_2))&=:\phi_{1\,\text{in}}(x_1)\phi_{2\,\text{in}}(x_2):+\langle 0|T(\phi_{1\,\text{in}}(x_1)\phi_{2\,\text{in}}(x_2))|0\rangle\\&=\phi_{2\,\text{in}}(x_2)\phi_{1\,\text{in}}(x_1)+\theta(x_{10}-x_{20})[\phi_{1\,\text{in}}(x_1),\phi_{2\,\text{in}}(x_2)].\end{aligned} \tag{148}$$

Note that the Green's function

$$G(x_1,x_2)=\langle 0|T(\phi_{1\,\text{in}}(x_1)\phi_{2\,\text{in}}(x_2))|0\rangle \tag{149}$$

is necessarily retarded. From this we can conclude that the models of electrodynamics, which we have considered, naturally embody the observed retarded nature of classical electrodynamics. Another way of stating this result is: If classical electrodynamics is to have a pseudoquantum formulation, its Green's functions are necessarily retarded. The origin of the asymmetry is the definition of the vacuums (which is equivalent to a specification of boundary conditions). Just as in classical electrodynamics retarded propagation is implemented by a choice of boundary conditions which do not require a commitment to any specific cosmological model.

Finally we would like to note that the Lagrangian obtained from adding L_{int} of Eq. (140) to the Lagrangian of Eq. (127) is equivalent to the usual Lagrangian of electrodynamics plus a term describing a massless Abelian gauge field with the wrong sign. (This is seen by defining new fields equal to the sum and difference of A^1_μ and A^2_μ.) This field theory may be quantized following the procedure we have outlined. A^1_μ loses its classical character due to quantum corrections.

IV. NON-ABELIAN GAUGE THEORIES

In this section we shall describe the procedure for embedding a classical non-Abelian Yang-Mills field in a quantum field theory. Then we will discuss a vierbein formulation of quantum gravity which could have been interpreted as a pseudoquantum field theory for a classical metric field if it were not for one term in the Lagrangian which makes it a truly quantum field theory. Nevertheless we suggest a new canonical quantization procedure based on our pseudoquantum approach.

Consider a classical Yang-Mills field, $A^1_\mu = \underline{A}^1_\mu \cdot \underline{T}$, where the jth component of $\underline{T}$ is a matrix representing a generator of a non-Abelian group G in the defining representation with commutation relations

$$[T_j, T_k] = i t_{jkl} T_l. \tag{150}$$

We can define a pseudoquantum field theory, wherein the classical character of A^1_μ is maintained, which has the Lagrangian density

$$\begin{aligned}\mathcal{L} = \tfrac{1}{2}\underline{F}^1_{\mu\nu}\cdot\underline{F}^{2\mu\nu} - \tfrac{1}{2}\underline{F}^{2\mu\nu}\cdot(\partial_\mu\underline{A}^1_\nu - \partial_\nu\underline{A}^1_\mu + g\underline{A}^1_\mu\times\underline{A}^1_\nu)\\ -\tfrac{1}{2}\underline{F}^{1\mu\nu}\cdot(\partial_\mu\underline{A}^2_\nu - \partial_\nu\underline{A}^2_\mu + g\underline{A}^1_\mu\times\underline{A}^2_\nu - g\underline{A}^1_\nu\times\underline{A}^2_\mu)\\ +\bar{\psi}(i\not\nabla + g\not A^1 - m)\psi,\end{aligned} \tag{151}$$

where ψ is a fermion field. The theory is invariant under the local gauge transformation, $S\in G$,

$$\psi' = S^{-1}\psi, \tag{152}$$

$$A^{1'}_\mu = S^{-1}A^1_\mu S + \frac{i}{g}S^{-1}\partial_\mu S, \tag{153}$$

$$F^{1'}_{\mu\nu} = S^{-1}F^1_{\mu\nu}S, \tag{154}$$

$$A^{2'}_\mu = S^{-1}A^2_\mu S, \tag{155}$$

$$F^{2'}_{\mu\nu} = S^{-1}F^2_{\mu\nu}S. \tag{156}$$

Except for one important term this Lagrangian with its attendant gauge invariance properties has been suggested as a possible model for the quark-confining strong interaction.[8] Since the omitted term has a masslike character $\Lambda^2\underline{A}^2_\mu\cdot\underline{A}^{2\mu}$, where Λ has the dimensions of a mass, it is clear that the strong-interaction model's ultraviolet behavior approaches that of the present pseudoquantum theory if the same quantization procedure is followed in both cases. We shall discuss this question further in the next section and show that the *ad hoc* procedure followed in Ref. 8 leads to the same result as the quantization procedure developed in this report.

The Euler-Lagrange equations of motion which are obtained from $\mathcal{L}$ in the canonical manner are

$$\underline{F}^1_{\mu\nu} = \partial_\mu\underline{A}^1_\nu - \partial_\nu\underline{A}^1_\mu + g\underline{A}^1_\mu\times\underline{A}^1_\nu, \tag{157}$$

$$\underline{F}^2_{\mu\nu} = \partial_\mu\underline{A}^2_\nu - \partial_\nu\underline{A}^2_\mu + g\underline{A}^1_\mu\times\underline{A}^2_\nu - g\underline{A}^1_\nu\times\underline{A}^2_\mu, \tag{158}$$

$$(\partial_\mu + g\underline{A}^1_\mu\times)\underline{F}^{1\mu\nu} = 0, \tag{159}$$

$$(\partial_\mu + g\underline{A}^1_\mu\times)\underline{F}^{2\mu\nu} + g\underline{A}^2_\mu\times\underline{F}^{1\mu\nu} + g\underline{J}^\nu = 0, \tag{160}$$

$$(i\overleftrightarrow{\nabla} + g\not A^1 - m)\psi = 0, \tag{161}$$

with the conservation law

$$(\partial_\nu + g\underline{A}^1_\nu\times)\underline{J}^\nu = 0. \tag{162}$$

The canonical momentum which is conjugate to $\underline{A}^1_j$ is

$$\underline{\Pi}^2_j = \underline{F}^2_{0j} \tag{163}$$

and the canonical momentum conjugate to $\underline{A}^2_j$ is

$$\underline{\Pi}^1_j = \underline{F}^1_{0j} \tag{164}$$

for $j = 1, 2, 3$. The canonical momentum corresponding to the fields $\underline{A}^i_0$ is zero for $i = 1, 2$. The existence of equations of constraint among the Euler-Lagrange equations implies that not all field components are independent, so that we must isolate the independent components prior to defining the canonical equal-time commutation relations.

Following Ref. 8 we choose to work in the Coulomb gauge, $\nabla_i A^1_i = 0$, and define the field variables

$$\underline{A}^2_i = \underline{A}^{2T}_i + \underline{A}^{2L}_i, \tag{165}$$

$$\underline{\Pi}^a_i = \underline{\Pi}^{aT}_i + \underline{\Pi}^{aL}_i, \tag{166}$$

where

$$\nabla_i\cdot\underline{A}^{2T}_i = \nabla_i\cdot\underline{\Pi}^{aT}_i = 0 \tag{167}$$

and $a = 1, 2$. Then the nonzero equal-time commutation relations are

$$[\Pi^{aT}_{ip}(x), A^{bT}_{jq}(y)] = i\delta_{pq}(1-\delta_{ab})\delta^{tr}_{ij}(\vec{x}-\vec{y}), \tag{168}$$

where p and q are internal-symmetry indices, $a,b = 1, 2$, and $i,j = 1, 2, 3$.

While the classical character of $\underline{A}^1_\mu$ can be maintained with our choice of $\mathcal{L}$, this theory has features due to its non-Abelian nature which make it less trivial and therefore more interesting than the corresponding Abelian theory discussed in the last section. If we follow a procedure similar to that in the Abelian case [Eq. (127)] and introduce a set of states appropriate to the quadratic part of the Lagrangian, then the cubic and quartic Yang-Mills terms in the interaction part of the Lagrangian will act to transform $\underline{A}^1_{\text{in}\,\mu}$ eigenstates into eigenstates of the interacting field $\underline{A}^1_\mu$. This is, of course, necessary for the classical Yang-Mills equations of motion to be satisfied. Our formalism, thus, offers a perturbative method for calculating solutions of the classical Yang-Mills equations. In addition, it gives an interesting interpretation to the short-distance behavior of the quark-confining field theory of Ref. 8. At short distances the gluon field $\underline{A}^1_\mu$ effectively decouples from the quark sector and becomes, in effect, a free field. This type of short-distance behavior is certainly not at odds with the seemingly simple behavior observed in hadron processes at high energy. Therefore, it is possible that pseudoquantum field theory may be relevant to the short-distance behavior of hadron interaction. Certainly, it is interesting that elementary fermions fall into two similar groups: those which appear to be individually observable (leptons) and those which are not individually observable (quarks).

We now turn to a consideration of a vierbein model of gravity which has certain close similarities to the pseudoquantum field theories we have been studying. In Weyl's formulation[13] of the Einstein-Cartan theory of gravity a vierbein field, $l^{\mu a}(x)$, is introduced which is the "square root" of the metric tensor

$$g^{\mu\nu}=\eta_{ab}l^{\mu a}l^{\nu b}, \tag{169}$$

where η_{ab} is the constant metric tensor of special relativity, where Roman indices transform as vectors under the SL(2, C) group of local Lorentz transformations, and where Greek indices transform as vectors under general coordinate transformations. It is useful to introduce the constant Dirac matrices, γ_a and $4S_{ab}=i[\gamma_a,\gamma_b]$. Under an SL(2, C) transformation,

$$S=\exp[iC^{ab}(x)S_{ab}], \tag{170}$$

a spinor, $\psi(x)$, becomes

$$\psi'=S\psi. \tag{171}$$

The local nature of the transformation requires the introduction of a gauge field

$$B^{ab}_\mu=-B^{ba}_\mu \tag{172}$$

which transforms inhomogeneously,

$$B_\mu\to SB_\mu S^{-1}-\frac{i}{g}S\partial_\mu S^{-1}, \tag{173}$$

so that a Lorentz transformation gauge-covariant derivative can be defined

$$\nabla_\mu\psi=(\partial_\mu+igB_\mu)\psi, \tag{174}$$

where $B_\mu=B^{ab}_\mu S_{ab}$ and $g=12\pi G$ where G is Newton's constant. Under a gauge transformation we have

$$l^\mu=l^{\mu a}\gamma_a\to Sl^\mu S^{-1}, \tag{175}$$

so that the gauge-covariant derivative of l^μ is defined to be

$$\nabla_\nu l^\mu=(\partial_\nu+igB_\nu\times)l^\mu, \tag{176}$$

where $B_\nu\times l^\mu=[B_\nu,l^\mu]$. The commutator

$$igB_{\mu\nu}=[\partial_\mu+igB_\mu,\partial_\nu+igB_\nu] \tag{177}$$

transforms homogeneously under a gauge transformation

$$B_{\mu\nu}\to SB_{\mu\nu}S^{-1}, \tag{178}$$

and as a second-rank tensor under general coordinate transformations. With these field quantities we are able to construct a Lagrangian $\mathcal{L}_{\text{Weyl}}$ which reduces to the Einstein Lagrangian for gravity when no matter is present,[13]

$$\mathcal{L}=\mathcal{L}_{\text{Weyl}}+\mathcal{L}_{\text{matter}}, \tag{179}$$

where

$$\mathcal{L}_{\text{Weyl}}=\frac{i}{8l}\operatorname{Tr}l^\mu l^\nu B_{\mu\nu} \tag{180}$$

and where, for example, we might let

$$l\,\mathcal{L}_{\text{matter}}=\bar\psi(il^\mu\nabla_\mu+m)\psi \tag{181}$$

with $l=\det(l^{\mu a})$.

We observe that the terms containing derivatives in $\mathcal{L}_{\text{Weyl}}$ are linear in the field B_μ—a suggestive feature in view of our previous discussion. However, the quadratic term in B_μ eliminates the possibility of regarding $\mathcal{L}_{\text{Weyl}}$ as a pseudoquantum field theory for a classical field $l^{\mu a}$. But, regardless of this consideration, the fact that $l^{\mu a}$ is necessarily classical in part leads us to consider quantizing vierbein gravity in a manner which is based on the pseudoquantization procedure described above. Remembering that a successful perturbation theory requires the perturbation to be around known solutions we introduce a quadratic Lagrangian term via

$$\mathcal{L}=\mathcal{L}_0+(\mathcal{L}-\mathcal{L}_0)=\mathcal{L}_0+\mathcal{L}_{\text{int}}, \tag{182}$$

where

$$\mathcal{L}_0=-\tfrac{1}{4}i\operatorname{Tr}(B'_{\mu a}l^\mu\gamma^a+ig[B_a,B_b]\gamma^a\gamma^b) \tag{183}$$

and

$$B'_{\mu a} = \partial_\mu B_a - \partial_a B_\mu \,. \tag{184}$$

Our plan is to follow the pseudoquantization procedure for the "free" part of the Lagrangian $\mathcal{L}_0$. Therefore we will (i) choose a particular coordinate system (harmonic coordinates) and a particular gauge, the "Lorentz" gauge, $\partial^\mu B_\mu = 0$, (ii) establish equal-time commutation relations, (iii) define a set of eigenstates of $l^{\mu a}$, and (iv) proceed to calculate quantum corrections in perturbation theory.

The equations of motion for the "free" Lagrangian $\mathcal{L}_0$ are

$$\partial_\mu B^{ab}_b - \partial_b B^{ab}_\mu = 0 \tag{185}$$

and

$$\partial_\mu (l^{\mu a}\eta^{\nu b} - l^{\nu a}\eta^{\mu b}) + 2g\,(\eta^{\nu a} B^{cb}_c - \eta^{\nu b} B^{ca}_c - \eta^{ac} B^{\nu b}_c + \eta^{bc} B^{\nu a}_c) = 0 \,. \tag{186}$$

We work in the gravitational equivalent of the Lorentz gauge of electrodynamics,

$$\partial^\mu B^{ab}_\mu = 0 \,, \tag{187}$$

and choose harmonic coordinates

$$\partial_\mu l^{\mu a} = \tfrac{1}{2}\, \partial^a \eta_{\sigma\tau} l^{\sigma\tau} \,. \tag{188}$$

The Green's function associated with Eq. (185) is

$$G_{\alpha ef,\rho\sigma}(x,y) = -\tfrac{1}{2} \int \frac{d^4k}{k^2}\, e^{-ik\cdot(x-y)} g_{\alpha ef,\rho\sigma}(k) \,, \tag{189}$$

where

$$g_{\alpha ef,\rho\sigma}(k) = k_e\left(\eta_{\alpha\rho}\eta_{f\sigma} + \eta_{\alpha\sigma}\eta_{f\rho} - \eta_{\alpha f}\eta_{\rho\sigma} - \frac{k_\alpha k_\rho \eta_{f\sigma} + k_\alpha k_\sigma \eta_{f\rho}}{k^2}\right) - k_f\left(\eta_{\alpha\rho}\eta_{e\sigma} + \eta_{\alpha\sigma}\eta_{e\rho} - \eta_{\alpha e}\eta_{\rho\sigma} - \frac{k_\alpha k_\rho \eta_{e\sigma} + k_\alpha k_\sigma \eta_{e\rho}}{k^2}\right). \tag{190}$$

In order to relate the above Green's function to a time-ordered product of the quantum fields it is first necessary to introduce a set of coherent states, $|L\rangle$, which are eigenstates of $l^{\mu a}$:

$$l^{\mu a}(x)\,|L\rangle = L^{\mu a}(x)\,|L\rangle \,, \tag{191}$$

where $L^{\mu a}(x)$ is a c-number function of x. In particular, we define $|\eta\rangle$ to satisfy

$$l^{\mu a}|\eta\rangle = \eta^{\mu a}|\eta\rangle \,, \tag{192}$$

where $\eta^{\mu a}$ is the constant Lorentz metric tensor of special relativity. Given a state $|L\rangle$ we define the field

$$l^{\mu a}_L = l^{\mu a} - L^{\mu a} \,. \tag{193}$$

This field corresponds to the quantum part of $l^{\mu a}$ and when applied to the purely classical state $|L\rangle$ has the eigenvalue zero.

We now make the identification

$$iG_{\alpha ef,\rho\sigma}(x,y) = \langle L\,|\,T(B_{\alpha ef}(x), l_{L\rho\sigma}(y))\,|\,L\rangle \,. \tag{194}$$

If we desire to calculate quantum corrections to $l_{\rho\sigma} = \eta_{\rho\sigma}$ we choose $|L\rangle = |\eta\rangle$. (It should be noted that $G_{\alpha ef,\rho\sigma}$ is independent of the choice of $|L\rangle$ as we have defined it.) Because $l_{L\rho\sigma}(y)$ is sharp on $|L\rangle$ we find that the right side of Eq. (194) becomes

$$iG_{\alpha ef,\rho\sigma}(x,y) = \theta(y_0 - x_0)[l_{\rho\sigma}(y), B_{\alpha ef}(x)] \tag{195}$$

up to a functional δ function. From the form of $\mathcal{L}_0$ we see that the commutator is not zero. It is fully determined by an equal-time commutation relation of $l_{\rho\sigma}$ and $B_{\alpha ef}$ (which by the way is the only nonzero equal-time commutator if the canonical procedure is followed), the equations of motion, and the requirement that it be zero at spacelike distances. The "retarded" form of $G_{\alpha ef,\rho\sigma}$ fixes the integration contour around poles in Eq. (192). The other nonzero Green's function in the free Lagrangian model specified by $\mathcal{L}_0$ is

$$iH^{\mu\nu,\rho\sigma}(x,y) = \langle L\,|\,T(l^{\mu\nu}_L(x), l^{\rho\sigma}_L(y))\,|\,L\rangle \,. \tag{196}$$

It is nonzero owing to the presence of the $[B_\mu, B_\nu]$ term in $\mathcal{L}_0$. We shall show in the next section that it is a principal-value propagator rather than a Feynman propagator. In coordinate space this results in $H^{\mu\nu,\rho\sigma}$ being the sum of the advanced and retarded propagators. As a result our model is equivalent to an action-at-a-distance theory in some sectors.

The classical part of $l_{\mu a}$ is the solution of the classical linearized field equations with appropriate matter sources. The linearized field equations are derived from a Lagrangian consisting of $\mathcal{L}_0$ plus matter terms. (Note that the form of $\mathcal{L}_0$ is obtained by substituting $l_{\mu a} = \eta_{\mu a} + h_{\mu a}$ in $\mathcal{L}_{\text{Weyl}}$, expanding, and keeping quadratic terms.) Thus the class of possible background metrics is restricted.

A simplification occurs in perturbation theory when the classical part of $l_{\mu\alpha}$ is $\eta_{\mu\alpha}$. In this case $(\mathcal{L}_{\text{Weyl}} - \mathcal{L}_0)|\eta\rangle = 0$ when $\mathcal{L}_0$ and $\mathcal{L}_{\text{Weyl}}$ are expressed in terms of asymptotic fields.

V. PRINCIPAL-VALUE PROPAGATORS AND ACTION AT A DISTANCE

In this section we shall show that certain propagators, in field theories where the pseudoquantization procedure has been followed, are principal-value propagators (i.e., the sum of the advanced and retarded Green's functions in coordinate space) rather than Feynman propagators. We also describe a quantum field theory for action-at-a-distance electrodynamics which completes the program initiated by Schwarzschild, Tetrode, and Fokker.[14]

To illustrate the origin of the principal-value propagator we return to the scalar field model of Eq. (103) which described a classical field, $\phi_1(x)$. We introduce an interaction term

$$L_{\text{int}} = -\int d^3z\, \tfrac{1}{2}\lambda^2 [\phi_2(z)]^2 \tag{197}$$

(where λ is a constant), which destroys the purely classical nature of ϕ_1. Suppose we consider the Green's function

$$i\bar{G}(x,y) = \langle 0|\, T(\phi_1(x)\phi_1(y))\,|0\rangle\,, \tag{198}$$

which would be zero if L_{int} were not present. In terms of in-fields we have

$$i\bar{G}(x,y) = \left\langle 0\left|T\left(\phi_{1\text{in}}(x)\phi_{1\text{in}}(y)\exp\left(i\int dt\, L_{\text{int}}\right)\right)\right|0\right\rangle, \tag{199}$$

where the vacuum states, $|0\rangle$ and $\langle 0|$, are defined as in Eqs. (120) and (122). From the definition of the vacuum we find (dropping "in" labels)

$$i\bar{G}(x,y) = \frac{-i\lambda^2}{2}\int d^4z \langle 0|\, T(\phi_1(x)\phi_1(y)\phi_2{}^2(z))\,|0\rangle\,, \tag{200}$$

which becomes

$$i\bar{G}(x,y) = \frac{-i\lambda^2}{2}\,\epsilon(x_0 - y_0)\,\frac{\partial}{\partial m^2}\,\Delta(x-y) \tag{201}$$

with

$$\Delta(x-y) = -i\int \frac{d^4k}{(2\pi)^3}\,\delta(k^2 - m^2)\epsilon(k_0)e^{-ik\cdot(x-y)}\,. \tag{202}$$

Using

$$\tfrac{1}{2}\epsilon(x_0 - y_0)\Delta(x-y) = \int \frac{d^4k}{(2\pi)^4}\,\text{P}\,\frac{1}{k^2 - m^2} \times e^{-ik\cdot(x-y)}\,, \tag{203}$$

we see that

$$\bar{G}(x,y) = -\lambda^2\int \frac{d^4k}{(2\pi)^4}\,\text{P}\,\frac{1}{(k^2-m^2)^2}\,e^{-ik\cdot(x-y)}\,, \tag{204}$$

where

$$\text{P}\,\frac{1}{(k^2-m^2)^2} \equiv \frac{1}{2}\left[\frac{1}{(k^2-m^2+i\epsilon)^2} + \frac{1}{(k^2-m^2-i\epsilon)^2}\right]. \tag{205}$$

The form of $\bar{G}$ is consistent with the equations of motion:

$$(\Box + m^2)\phi_1 + \lambda^2\phi_2 = 0\,, \tag{206}$$

$$(\Box + m^2)\phi_2 = \delta^4(x-y)\,. \tag{207}$$

The appearance of the principal-value dipole propagator rather than the Feynman dipole propagator in Eq. (204) is useful because it eliminates certain unitarity problems associated with indefinite-metric fields. However, depending on the model under consideration, it could lead to difficulties with causality. To illustrate the manner in which unitarity problems are resolved, consider the interaction of the ϕ_1 dipole field with a scalar quantum field ψ with

$$L'_{\text{int}} = g\phi_1(x)[\psi(x)]^2\,. \tag{208}$$

Suppose we consider the subset of in and out states containing arbitrary numbers of ψ particles but no ϕ_1 or ϕ_2 particles. These states have positive metric. If one could systematically exclude indefinite-metric ϕ_1 and ϕ_2 particles from physical states one would avoid negative probabilities and other problems. But the sum over states in a unitarity sum would normally include states with ϕ_1 particles if the ϕ_1 field had Feynman propagators. In the case of principal-value propagators, no intermediate states with ϕ_1 particles occur, since the pole term is not present. The interaction mediated by the ϕ_1 field is a form of action at a distance and ϕ_1 is properly described by the phrase adjunct field, coined by Feynman and Wheeler.[14] A more detailed discussion of the unitarity question is given in Refs. 7 and 8. In those articles a dipole gluon model for quark confinement was proposed which introduced principal-value propagators in an *ad hoc* manner to resolve unitarity problems. It was pointed out that causality problems did not necessarily exist in those models because the non-Abelian dipole gluons were confined for the same reason as the quarks so that—at the worst—there would be unobservable causality violations at distances of the order of hadron dimensions.

The pseudoquantization procedure may be used to construct a quantum field-theoretic version of action-at-a-distance electrodynamics. Consider the Lagrangian

$$\mathcal{L} = -\tfrac{1}{2}F^{\mu\nu}(\partial_\nu A_\mu - \partial_\mu A_\nu) + \tfrac{1}{4}F^{\mu\nu}F_{\mu\nu} + \bar{\psi}(i\not{\partial} - e\not{A} - m_0)\psi\,. \tag{209}$$

We define the momentum

$$\Pi_\mu = \frac{\delta \mathfrak{L}}{\delta \dot{A}^\mu} = F_{0\mu} . \tag{210}$$

Going to the transverse gauge as in Sec. IV, we define the equal-time commutation relation

$$[\Pi_i(\vec{x}, t), A_j(\vec{y}, t)] = i\delta_{ij}^{tr}(\vec{x} - \vec{y}) . \tag{211}$$

Suppose we neglect interaction terms in $\mathfrak{L}$ for the moment and choose $F_{\mu\nu}$ to be an observable classical field (as it is up to quantum corrections which we neglect) and A_μ to be unobservable (as it is because it is not gauge invariant). Then we follow our pseudoquantization procedure for

$$\mathfrak{L}_0 = -\tfrac{1}{2} F^{\mu\nu}(\partial_\nu A_\mu - \partial_\mu A_\nu) + \tfrac{1}{4} F^{\mu\nu} F_{\mu\nu} . \tag{212}$$

In particular, we define a vacuum such that

$$F_{\mu\nu}|0\rangle = 0, \quad A_\mu|0\rangle \neq 0 , \tag{213}$$

while

$$\langle 0|A_\mu = 0, \quad \langle 0|F_{\mu\nu} \neq 0 . \tag{214}$$

Then

$$iG_{\mu\nu}(x, y) = \langle 0|T(A_\mu(x)A_\nu(y))|0\rangle \tag{215}$$

would be zero were it not for $F_{\mu\nu}F^{\mu\nu}$ in $\mathfrak{L}_0$. In terms of appropriate in-fields it becomes

$$\begin{aligned} 2iG_{\mu\nu}(x, y) = \int d^4z\, (&\theta(x_0 - y_0)\theta(y_0 - z_0) \\ &+ \theta(y_0 - x_0)\theta(x_0 - z_0)) \\ &\times [A_{\mu\,\text{in}}(x), F_{\alpha\beta\,\text{in}}(z)][A_{\mu\,\text{in}}(y), F_{\text{in}}^{\alpha\beta}(z)] . \end{aligned} \tag{216}$$

Note that we are treating $F_{\mu\nu}F^{\mu\nu}$ in $\mathfrak{L}_0$ as an interaction term. The structure of $G_{\mu\nu}(x, y)$ is the same as that of Eq. (200) so we can conclude that

$$G_{\mu\nu}(x, y) = -g_{\mu\nu} \int \frac{d^4k}{(2\pi)^4}\, \mathrm{P}\frac{1}{k^2}\, e^{-ik\cdot(x-y)} \tag{217}$$

in the Feynman gauge. Thus the action-at-a-distance interaction follows from the pseudoquantization of electrodynamics. The classical character of $F_{\mu\nu}$ is lost owing to quantum corrections resulting from the presence of $J_\mu A^\mu$ in the Lagrangian.

The example we have just studied has a certain parallel in the vierbein model of gravitation studied in the last section. The forms of the Lagrangian and commutation relations are similar. As a result it is clear that

$$D^{\mu\nu,\lambda\sigma}(x, y) \equiv \left\langle L \left| T\left(l_{L\,\text{in}}^{\mu\nu}(x) l_{L\,\text{in}}^{\lambda\sigma}(y) \int d^4z\, \tilde{\mathfrak{L}}_{\text{int}}(z)\right) \right| L \right\rangle \tag{218}$$

with

$$\tilde{\mathfrak{L}}_{\text{int}} = \tfrac{1}{4} g \operatorname{Tr}[B_{\mu\,\text{in}}, B_{\nu\,\text{in}}]\gamma^\mu\gamma^\nu \tag{219}$$

is a principal-value propagator. Therefore we have constructed an action-at-a-distance version of quantum gravity. Our motivation was to take account of the classical part of $l^{\mu a}$ in a way which did not divorce it from the quantum part to which it is intimately related.

VI. CONCLUSION

We have seen that an alternative to Fock-space quantization exists for a class of field theories which have Lagrangian gradient terms which are linear in field variables. A method was also proposed for constructing Lagrangians of that type from classical Lagrangians with gradient terms which are quadratic in field variables. To some extent this process has a parallel in the passage from Klein-Gordon field Lagrangians which are quadratic in derivatives to Dirac field Lagrangians which are linear in derivatives.

The quantization procedure we have outlined is canonical so far as the fields are concerned. We do, however, make a choice of vacuum states which differs from the usual choice. As a result we have found free propagators which were either retarded, or half-advanced and half-retarded. The choice of vacuum state does not in itself preclude the appearance of Feynman propagators. If one has a good reason to modify the canonical commutation relations then it is possible to obtain Feynman propagators.[15] The procedure we have outlined has, therefore, a greater generality than the particular class of models studied in the present work. It can enable one to embed a classical field theory in a quantum field theory in such a way as to maintain its classical character. It can also be applied to study classical field theories which obtain quantum corrections. Finally it can be applied in order to obtain a fully second-quantized field theory (cf. Ref. 15).

ACKNOWLEDGMENT

This work was supported in part by the U.S. Energy Research and Development Administration.

---

*Present address: Physics Department, Williams College, Williamstown, Mass. 01267.

[1]D. R. Yennie, S. C. Frautschi, and H. Suura, Ann. Phys. (N.Y.) 13, 379 (1961).

[2]R. J. Glauber, Phys. Rev. 131, 2766 (1963).

[3]W. A. Bardeen, M. S. Chanowitz, S. D. Drell, M. Weinstein, and T.-M. Yan, Phys. Rev. D 11, 1094 (1975).

[4]J. M. Cornwall and G. Tiktopoulos, Phys. Rev. D 13, 3370 (1976).
[5]S. Blaha, Phys. Lett. 56B, 373 (1975).
[6]E. C. G. Sudarshan, Center for Particle Theory report Univ. of Texas—Austin, 1976 (unpublished).
[7]S. Blaha, Phys. Rev. D 10, 4268 (1974).
[8]S. Blaha, Phys. Rev. D 11, 2921 (1975).
[9]S. Blaha, Lett. Nuovo Cimento 18, 60 (1977).
[10]Cf. Ref. 2; T. W. B. Kibble, J. Math. Phys. 9, 315 (1968); Phys. Rev. 173, 1527 (1968); 174, 1882 (1968); 175, 1624 (1968);
[11]A. Salam, lecture at Center for Theoretical Studies, Miami, Florida, 1973 (unpublished).
[12]G. Mie, Ann. Phys. (Leipzig) 37, 511 (1912); 39, 1 (1912); 40, 1 (1913); H. Weyl, *Space, Time, Matter* (Dover, N.Y. 1952).
[13]H. Weyl, Z. Phys. 56, 330 (1929); T. W. B. Kibble, J. Math. Phys. 2, 212 (1961); J. Schwinger, Phys. Rev. 130, 1253 (1963); C. J. Isham, A. Salam, and J. Strathdee, Lett. Nuovo Cimento 5, 969 (1972); F. W. Hehl, P. von der Heyde, G. D. Kerlick, and J. Nester, Rev. Mod. Phys. 48, 393 (1976); and references therein.
[14]K. Schwarzschild, Göttinger Nachrichten 128, 132 (1903); H. Tetrode, Z. Phys. 10, 317 (1922); A. D. Fokker, *ibid.* 58, 386 (1929); J. Wheeler and R. P. Feynman, Rev. Mod. Phys. 17, 157 (1945); 21, 425 (1949).
[15]S. Blaha (unpublished).

Appendix D. Unified Theory of Gravitation and the Strong Interaction

The following essay, by this author, is a copy of the original essay awarded Honorable Mention by the Gravity Research Foundation Essay Competition.

C00-3533-68
SU-4208-68
March 1976

QUANTUM GRAVITY AND QUARK CONFINEMENT*

Stephen Blaha
Department of PHysics
Syracuse University, Syracuse, New York 13210

ABSTRACT

We point out certain formal similarities between gravity and field theoretic models of the strong interaction which realize quark confinement through the Schwinger mechanism. A unified theory of the strong interaction and gravity is constructed based on the requirements of (i) manifest renormalizability and quark confinement in perturbation theory, (ii) simplicity in the choice of a mechanism to realize (i), and (iii) a symmetric coordination of the SL(2,C) gauge fields in gravity, and the color SU(3) Yang-Mills fields embodying the strong interaction so that quark confinement and renormalizability of quantum gravity are correlated.

*Work supported in part by the U.S. Energy Research and Development Administration (ERDA)

The Weinberg-Salam models of the weak and electromagnetic interactions show that considerations of symmetry and renormalizability can outweigh arguments against unified field theories based on widely differing interaction strengths. Symmetry considerations have even formed the basis of attempts to unite the strong interactions and gravity.[1,2]

We shall develop a unified theory of gravity and the strong interaction which is motivated by some formal similarities between gravity and field theory models in which quark confinement follows from the Schwinger mechanism. We shall assume that the failure to isolate a quark reflects a fundamental fact of nature: that hadrons cannot be subdivided into their quark constituents. This property can be realized in a field theory [3,4,5] where the quark-quark strong interaction is mediated by massless, colored, Yang-Mills fields which acquire a mass through vacuum polarization effects (the Schwinger mechanism) while maintaining the unbroken nature of the color symmetry. If the Yang-Mills color symmetry group is SU(3) then physical states are either quark-anti-quark states (mesons) or three quark states (baryons) - in agreement with experiment.

A common property of gravity and gauge theories with Schwinger mechanism is apparent in the first purely field theoretic model of quark confinement, two-dimensional quantum electrodynamics.[3] In gravity, arbitrary additive changes in the scale of the energy, $E \to E+c$, are not allowed because the energy-momentum tensor is the source of the gravitational field. In Schwinger mechanism gauge theories (abelian) such changes are not admitted because the gauge group does not include transformations of the type, $\Lambda=ct$, where t is the time[6]. As a result the possibility of compensating shifts in the energy by a gauge transformation of the potential, $A_0 \rightarrow A_0+c/e$, is lost. The restriction of the gauge group due to the Schwinger mechanism reflects its close relation to spontaneous breakdown. Spontaneous breakdown, of course, occurs in the background field formulation of quantum gravity.

The unbroken nature of the color symmetry, a property which is crucial for the realization of quark confinement via the Schwinger mechanism, provides another point of similarity with gravity. The color symmetry gauge group stands with the group of general coordinate transformations as the only known exact non-abelian transformation groups. This leads one to suspect that the dynamics of color may have an intimate relationship with the geometry of the universe. One possible clue to this relationship may be in the observation that a requirement for a complete field theory of quantum gravity, renormalizability, and an apparently major feature of the strong interaction, quark confinement through the Schwinger mechanism, are both attainable through the same device: the introduction of fourth order derivatives in the field equations in the respective cases.[4,7] In a sense the introduction of higher order derivatives is the simplest innovation which manifestly leads to these properties in perturbation theory - the only generally effective method of computation available. There are no decisive arguments against such terms based on unitarity since a number of methods are now available for unitarizing field theories with higher derivative terms.[4,8,9]

Taking these similarities between gravity and the quark confining strong interaction as a guide we can construct a model unified field theory of the strong interaction and gravity. For simplicity we neglect other interactions. In view of the spin 1/2 nature of quarks our starting point is Weyl's formulation of the Einstein-Cartan (ECW) theory of gravity.[1,10] We introduce the vierbein field, $l^{\mu a}(x)$, and the spinor connection, $B_{\mu ab}(x)$, with the metric tensor $g^{\mu\nu} = \eta_{ab} l^{\mu a} l^{\nu b}$. Under a local Lorentz transformation, S, of the SL(2,C) gauge group we find

$$\psi \to S\psi \tag{1}$$

$$l^{\mu} \to S\, l^{\mu} S^{-1} \tag{2}$$

$$B^{\mu} \to S B^{\mu} S^{-1} - i S \partial_{\mu} S^{-1} \tag{3}$$

$$(\partial_{\mu} + i B_{\mu})\psi \to S(\partial_{\mu} + i B_{\mu})\psi \tag{4}$$

where ψ is a spinor field, $l^{\mu} = l^{\mu a}\gamma_a$, $B_{\mu} = B_{\mu ab} S^{ab}$ and $4 S^{ab} = i[\gamma^a, \gamma^b]$. The affine connection is introduced by defining the covariant derivative of l_{μ} to be zero:

$$D_{\nu} l_{\mu} \equiv \partial_{\nu} l_{\mu} + i[B_{\nu}, l_{\mu}] - \binom{\sigma}{\nu\mu} l_{\sigma} = 0 \tag{5}$$

where l_{μ} is inver[illegible] of l^{μ}.

Eq.(5) may be rewritten in the form:

$$D_{\nu}\ell_{\mu} = \partial_{\nu}\ell_{\mu} + i[B_{\nu}, \ell_{\mu}] + i[\tilde{B}^{\sigma}_{\nu\mu}, \ell_{\sigma}] = 0 \tag{6}$$

where $\tilde{B}^{\sigma}_{\nu\mu} = \tilde{B}^{\sigma ab}_{\nu\mu} S_{ab}, -2i\,\tilde{B}^{\alpha ab}_{\nu\mu}\ell_{\alpha a}\ell^{\sigma}_{b} = \binom{\sigma}{\nu\mu}$ and $\tilde{B}^{\sigma ab}_{\nu\mu}$ transforms homogeneously under the SL(2,C) gauge group. Our purpose in introducing $\tilde{B}^{\sigma ab}_{\nu\mu}$ is to have additional field variables at our disposal, in an algebraically constrained manner, with a view towards using them to obtain fourth order field equations, within the framework of the canonical second order formalism, through the Ostrogradski bootstrap.[11] (The usual ECW Lagrangian, in terms of ℓ^{μ} and B_{ν}, is a first order formalism which realizes second order equations of motion through the Ostrogradski bootstrap.) We now generalize our field variables to include colorful fields: $\tilde{\ell}_{\mu} = \tilde{\ell}^{ai}_{\mu}\gamma_a T^i$ where, for i=1,2,...,8, $\tilde{\ell}^{ai}_{\mu}$ transforms according to the adjoint representation of SU(3) and, for i=9, according to the scalar (singlet) representation, with $\ell^{\mu a} = \tilde{\ell}^{\mu a 9}$. In analogy to B_{μ} and $\tilde{B}^{\sigma}_{\nu\mu}$ we introduce the Yang-Mills fields, A_{μ} and $\tilde{A}^{\sigma}_{\nu\mu}$, which are general coordinate transformation tensors, SL(2,C) scalars, and under an SU(3) gauge transformation, C:

$$\tilde{\ell}^{\mu} \longrightarrow C\tilde{\ell}^{\mu}C^{-1} \tag{7}$$

$$A_{\mu} \longrightarrow CA_{\mu}C^{-1} - iC\partial_{\mu}C^{-1} \tag{8}$$

$$\tilde{A}^{\sigma}_{\nu\mu} \longrightarrow C\tilde{A}^{\sigma}_{\nu\mu}C^{-1} \tag{9}$$

We <u>define</u> a covariant derivative

$$D_\mu V_\nu \equiv (\partial_\mu + iB_\mu\times + iA_\mu\times)V_\nu + i(\tilde{B}^\sigma_{\mu\nu} + \tilde{A}^\sigma_{\mu\nu})\times V_\sigma \tag{10}$$

where E×F=EF-FE, and a curvature tensor by $(D_\rho D_\sigma - D_\sigma D_\rho)V_\nu = \tilde{R}^\beta_{\nu\rho\sigma} V_\beta$

For the purposes of the Ostrogradski bootstrap it suffices to let

$\tilde{B}^\sigma_{\nu\mu} = \frac{1}{5}(g^\sigma_\nu \tilde{B}_\mu + g^\sigma_\mu \tilde{B}_\nu) + \tilde{\tilde{B}}^\sigma_{\nu\mu}$ and $5\tilde{A}^\sigma_{\nu\mu} = g^\sigma_\nu \tilde{A}_\mu + g^\sigma_\mu \tilde{A}_\nu$

where $\tilde{B}_\mu$ and $\tilde{A}_\mu$ are vectors under general coordinate transformations. We then extract from $\tilde{R}^\beta_{\nu\rho\sigma}$ the gauge covariant (but non-tensorial quantity:

$$\begin{aligned} R^\beta_{\nu\rho\sigma} = {} & i g^\beta_\nu \left(F^1_{\rho\sigma} + B^1_{\rho\sigma} - \frac{1}{5}F^2_{\rho\sigma} - \frac{1}{5}B^2_{\rho\sigma} - \frac{i}{25}\tilde{B}_\sigma\times\tilde{B}_\rho - \frac{i}{25}\tilde{A}_\sigma\times\tilde{A}_\rho\right) - \\ & - \frac{i}{5} g^\beta_\sigma (\partial_\rho + iB_\rho\times)\tilde{B}_\nu + \frac{i}{5} g^\beta_\rho (\partial_\sigma + iB_\sigma\times)\tilde{B}_\nu - \frac{i}{5} g^\beta_\sigma (\partial_\rho + iA_\rho\times)\tilde{A}_\nu \\ & + \frac{i}{5} g^\beta_\rho (\partial_\sigma + iA_\sigma\times)\tilde{A}_\nu + \frac{1}{25}\left[g^\beta_\rho(\tilde{B}_\sigma\tilde{B}_\nu + \tilde{A}_\sigma\tilde{A}_\nu) - \right. \\ & \left. - g^\beta_\sigma(\tilde{B}_\rho\tilde{B}_\nu + \tilde{A}_\rho\tilde{A}_\nu)\right] \end{aligned} \tag{11}$$

and define the "Ricci" contraction, $R_{\mu\nu} = R^\beta_{\mu\nu\beta}$, which has the antisymmetric tensorial part

$$R^A_{\mu\nu} = \frac{1}{2}(R_{\mu\nu} - R_{\nu\mu}) = -i(F^1_{\mu\nu} + B^1_{\mu\nu}) + \frac{i}{2}(F^2_{\mu\nu} + B^2_{\mu\nu}) + \frac{1}{10}(\tilde{B}_\mu\times\tilde{B}_\nu + \tilde{A}_\mu\times\tilde{A}_\nu) \tag{12}$$

where

$$F^1_{\mu\nu} = \partial_\mu A_\nu - \partial_\nu A_\mu + i A_\mu \times A_\nu \tag{13}$$

$$F^2_{\mu\nu} = \partial_\mu \tilde{A}_\nu - \partial_\nu \tilde{A}_\mu + i A_\mu \times \tilde{A}_\nu - i A_\nu \times \tilde{A}_\mu \tag{14}$$

$$B^1_{\mu\nu} = \partial_\mu B_\nu - \partial_\nu B_\mu + i B_\mu \times B_\nu \tag{15}$$

$$B^2_{\mu\nu} = \partial_\mu \tilde{B}_\nu - \partial_\nu \tilde{B}_\mu + i B_\mu \times \tilde{B}_\nu - i B_\nu \times \tilde{B}_\mu \tag{16}$$

Another non-trivial antisymmetric tensor is

$$S_{\mu\nu} \equiv R^{\beta}_{\ \beta\mu\nu} = 4i\left(F^1_{\mu\nu} + B^1_{\mu\nu}\right) - i\left(B^2_{\mu\nu} + F^2_{\mu\nu}\right) - \frac{1}{5}\left(\tilde{B}_\mu \times \tilde{B}_\nu + \tilde{A}_\mu \times \tilde{A}_\nu\right) \tag{17}$$

An examination of eqs.(11)-(17) shows that they are invariant under the interchange: $A_\mu \leftrightarrow B_\mu$ and $\tilde{A}_\mu \leftrightarrow \tilde{B}_\mu$. We have imposed this symmetry which we call gauge field coordination in order to restrict the form of the tensorial quantities and to partially correlate the dynamics of gravity and the strong interaction. The Lagrangian we construct from the curvature tensor and its contractions is

$$l\mathcal{L} = \mathrm{Tr}\left\{\frac{-1}{24f^2}\left(S^{\mu\nu} + 2R^{A\mu\nu}\right)\left(S_{\mu\nu} + 4R^A_{\mu\nu}\right) + \frac{1}{6K} l^\mu l^\nu \left(R_{\mu\nu} + \frac{1}{2} S_{\mu\nu}\right)\right\} \tag{18}$$

where $l = d_o + l^{\mu a}$, $K = 8\pi G$, f is a dimensionless coupling constant, and only the ninth component of $\tilde{l}_\mu$ appears in $\mathcal{L}$ for simplicity.

Because of the vast disparity between the strengths of the strong interaction and gravity we cannot expect to see detailed experimental confirmation of a model of the type of eq.(17) through a study of the interplay of the two forces. Rather we would expect such a model to be acceptable if it provides (i) an, at first, qualitative and eventually a quantitative description of strong interaction phenomena, (ii) a large distance theory of gravitation in accord with experiment, (iii) a short distance, renormalizable field theory of quantum gravity, and finally, (iv) a natural starting point for the incorporation of the remaining weak and electromagnetic interactions, in an essential way in a fully unified theory (rather than an ad hoc addition of terms to the Lagrangian). Points (ii) and (iii) are explicitly realized in our model. The strong interaction sector of our model is virtually identical to a previously studied model[4] of the strong interactions. As a result we can conclude that the strong interaction sector of our model with effective Lagrangian:

$$\mathcal{L}_{STRONG} = \frac{1}{2f^2} F^{1\mu\nu} \cdot \left(F^2_{\mu\nu} - \frac{i}{5} \tilde{A}_\mu \times \tilde{A}_\nu \right) - \frac{3}{25K} \tilde{A}^\mu \cdot \tilde{A}_\mu \qquad (19)$$

exhibits quark confinement, Bjorken scaling in the deep inelastic lepton-nucleon scattering structure functions, and a linear confining effective potential, $V \propto |\vec{r}|$ which appears to successfully describe features of the new ψ/J heavy mesons.[12] Thus there are strong indications that eq.(19) provides a reasonable model of the strong interaction on the qualitative level. It is interesting to note that the scale of the strong interactions is set by Newton's constant, G, in our model.

The remaining test for the unified field, (iv), may also be passed.

First, we note, with Kibble,[10] that in the ECW model there is a surprising similarity in form of the effective spin-spin interaction, $K(\bar{\psi}\gamma_\mu\gamma_5\psi)^2$ to Fermi weak interaction terms. The similarity, in our case, is actually to terms occurring in intermediate vector boson models of the weak interaction such as the Weinberg-Salam models. Of course, an internal symmetry group of the type observed in the weak interactions does not appear in our model. We might conjecture that such a symmetry could arise from a factorization in the dynomical equations analogous to the factorization of the Klein-Gordon equation which led to the Dirac matrix algebra. A hint of this possibility is contained in a numerical coincidence. Comparing eq.(18) with the Lagrangian of reference 4, we find that a dimensionless strong interaction coupling constant, f, and a constant with dimensions of mass^2, $\lambda^2 = 6f^2/(25K)$ are defined and that as a result perturbation theory is an expansion in the quantity, $f^2\lambda^2/\pi$, where π enters from loop integrations for geometrical reasons. For momenta much greater than this quantity the theory is quasi-free and, in the case of deep inelastic structure functions, Bjorken scaling follows. Since Bjorken scaling is evident at momenta $\sim 1\ \text{GeV}^2$ we can take $f^2\lambda^2 \cong \pi\ \text{GeV}^2$ and use our expression for λ^2 in terms of K to calculate $\lambda^2 = 2.12\cdot 10^{18}\ \text{GeV}^2$ and $f^2 = 1.48\cdot 10^{-18}$, Then $\sqrt{f} = 3.49\cdot 10^{-5}$ approximately equals the weak decay constant, $G_{Fermi} = 3.61\ 10^{-5}$ measured in units of $f^2\lambda^2$, i.e. $\sqrt{f} \cong f^2\lambda^2 G_{FERMI}$ where $G_{Fermi} \cong 1.15\ \text{GeV}^{-2}$. The fact that $\sqrt{f}$, rather than f or f^2, is equal to $G_{Fermi}f^2\lambda^2$ lends support to the factorization conjecture.

We have shown that there are some grounds for expecting a unified theory of the known interactions and have constructed a simple model which embodies major physical requirements in a compact and symmetric manner.

REFERENCES

1. C.Isham, A.Salam and J.Strathdee, Phys. Rev. D3, 867 (1971); B.Zumino, Lectures on Elementary Particles and Quantum Field Theory, M.I.T. Press, Cambridge, Mass. (1970), vol.II.
2. J.Wess and B.Zumino, Nucl. Phys. B70, 39 (1974); D.V.Volkov and V.Soroka, JETP Lett. 18, 529 (1973); A.Salam and J.Strathdee, Nucl. Phys. B76, 477 (1974); R.Arnowitt and P.Nath, Northeastern Univ. Preprint NUB 2261 (1975).
3. A.Casher, J.Kogut and L.Susskind, Phys. Rev. D10, 732 (1974); J.Lowenstein and J.Swieca, Ann. Phys. (N.Y.) 68, 172 (1971).
4. S.Blaha, Phys. Rev. D11, 2921 (1975).
5. L.-F.Li and J.Willemsen, Phys. Rev. D10, 4087 (1974).
6. R.Brandt and Ng Wing-Chiu, Phys. Rev. D10, 4198 (1974).
7. S.Deser, H.-S.Tsao, and P.van Nieuwenhuizen, Phys. Rev. D10, 3337 (1974).
8. E.C.G.Sudarshan, Fields and Quanta 2, 175 (1972); M.Karowski, Nuov. Cim. 24A, 126 (1974); K.L.Nagy, Acta Phys. Hung., 32, 127 (1972).
9. K.L.Nagy, Eötvös Univ. Preprint 350 (1975).
10. H.Weyl, E. Physik 56, 330 (1929); T.W.B.Kibble, J.Math Phys. 2, 212 (1961); J.Schwinger, Phys. Rev. 130, 1253 (1963); C.J.Isham, A.Salam and J.Strathdee, Lettere al Nuovo Cimento 5, 969 (1972); and references therein.
11. M.Ostrogradski, Mem. Ac. St.Petersbourg VI 4, 385 (1850); A.Pais and G.Uhlenbeck, Phys. Rev. 79, 145 (1950).
12. E.Eichten et al, Phys. Rev. Lett. 34, 52 (1975).

REFERENCES

Bjorken, J. D., Drell, S. D., 1964, *Relativistic Quantum Mechanics* (McGraw-Hill, New York, 1965).

Bjorken, J. D., Drell, S. D., 1965, *Relativistic Quantum Fields* (McGraw-Hill, New York, 1965).

Blaha, S., 1998, *Cosmos and Consciousness* (Pingree-Hill Publishing, Auburn, NH, 1998).

_______, 2002, *A Finite Unified Quantum Field Theory of the Elementary Particle Standard Model and Quantum Gravity Based on New Quantum Dimensions™ & a New Paradigm in the Calculus of Variations* (Pingree-Hill Publishing, Auburn, NH, 2002).

_______, 2003, *A Finite Unified Quantum Field Theory of the Elementary Particle Standard Model and Quantum Gravity Based on New Quantum Dimensions™ and a New Paradigm in the Calculus of Variations* (Pingree-Hill Publishing, Auburn, NH, 2003).

_______, 2004, *Quantum Big Bang Cosmology: Complex Space-time General Relativity, Quantum Coordinates™ Dodecahedral Universe, Inflation, and New Spin 0, ½, 1 & 2 Tachyons & Imagyons* (Pingree-Hill Publishing, Auburn, NH, 2004).

______, *2005a, Quantum Theory of the Third Kind: A New Type of Divergence-free Quantum Field Theory Supporting a Unified Standard Model of Elementary Particles and Quantum Gravity based on a New Method in the Calculus of Variations* (Pingree-Hill Publishing, Auburn, NH, 2005).

______, 2005b, *The Metatheory of Physics Theories, and the Theory of Everything as a Quantum Computer Language* (Pingree-Hill Publishing, Auburn, NH, 2005).

______, 2005c, *The Equivalence of Elementary Particle Theories and Computer Languages: Quantum Computers, Turing Machines, Standard Model, Superstring Theory, and a Proof that Gödel's Theorem Implies Nature Must Be Quantum* (Pingree-Hill Publishing, Auburn, NH, 2005).

______, 2006a, *The Foundation of the Forces of Nature* (Pingree-Hill Publishing, Auburn, NH, 2006).

______, 2006b, *A Derivation of ElectroWeak Theory based on an Extension of Special Relativity; Black Hole Tachyons; & Tachyons of Any Spin*. (Pingree-Hill Publishing, Auburn, NH, 2006).

______, 2007a, *Physics Beyond the Light Barrier: The Source of Parity Violation, Tachyons, and A Derivation of Standard Model Features* (Pingree-Hill Publishing, Auburn, NH, 2007).

______, 2007b, *The Origin of the Standard Model: The Genesis of Four Quark and Lepton Species, Parity Violation, the ElectroWeak Sector, Color SU(3), Three Visible Generations of Fermions, and One Generation of Dark Matter with Dark Energy* (Pingree-Hill Publishing, Auburn, NH, 2007).

______, *2008a, A Direct Derivation of the Form of the Standard Model From GL(16) (Pingree-Hill Publishing, Auburn, NH, 2008).*

______, 2008b, *A Complete Derivation of the Form of the Standard Model With a New Method to Generate Particle Masses Second Edition* (Pingree-Hill Publishing, Auburn, NH, 2008)

______, 2009, *The Algebra of Thought & Reality: The Mathematical Basis for Plato's Theory of Ideas, and Reality Extended to Include A Priori Observers and Space-Time Second Edition* (Pingree-Hill Publishing, Auburn, NH, 2009).

______, 2010a, *Operator Metaphysics: A New Metaphysics Based on a New Operator Logic and a New Quantum Operator Logic that Lead to a Mathematical Basis for Plato's Theory of Ideas and Reality* (Pingree-Hill Publishing, Auburn, NH, 2010).

______, 2010b, *The Standard Model's Form Derived from Operator Logic, Superluminal Transformations and GL(16)* (Pingree-Hill Publishing, Auburn, NH, 2010).

______, 2011a, *21st Century Natural Philosophy Of Ultimate Physical Reality* (McMann-Fisher Publishing, Auburn, NH, 2011).

______, 2011b, *All the Universe! Faster Than Light Tachyon Quark Starships & Particle Accelerators with the LHC as a Prototype Starship Drive Scientific Edition* (Pingree-Hill Publishing, Auburn, NH, 2011).

______, 2011c, *From Asynchronous Logic to The Standard Model to Superflight to the Stars* (Blaha Research, Auburn, NH, 2011).

______, 2012a, *From Asynchronous Logic to The Standard Model to Superflight to the Stars volume 2: Superluminal CP and CPT, U(4) Complex General Relativity and The Standard Model, Complex Vierbein General Relativity, Kinetic Theory, Thermodynamics* (Blaha Research, Auburn, NH, 2012).

______, 2012b, *Standard Model Symmetries, And Four And Sixteen Dimension Complex Relativity; The Origin Of Higgs Mass Terms* (Blaha Reasearch, Auburn, NH, 2012).

______, 2013a, *Multi-Stage Space Guns, Micro-Pulse Nuclear Rockets, and Faster-Than-Light Quark-Gluon Ion Drive Starships* (Blaha Research, Auburn, NH, 2013).

______, 2013b, *The Bridge to Dark Matter; A New Sister Universe; Dark Energy; Inflatons; Quantum Big Bang; Superluminal Physics; An Extended Standard Model Based on Geometry* (Blaha Reasearch, Auburn, NH, 2013).

______, 2014a, *Universes and Multiverses: From a New Standard Model to a Physical Multiverse; The Big Bang; Our Sister Universe's Wormhole; Origin of the Cosmological Constant, Spatial Asymmetry of the Universe, and its Web of Galaxies; A Baryonic Field between Universes and Particles; Flatverse Extended Wheeler-DeWitt Equation* (Blaha Reasearch, Auburn, NH, 2014).

______, 2014b, *All the Multiverse! Starships Exploring the Endless Universes of the Cosmos Using the Baryonic Force* (Blaha Research, Auburn, NH, 2014).

______, 2014c, *All the Multiverse! II Between Multiverse Universes: Quantum Entanglement Explained by the Multiverse Coherent Baryonic Radiation Devices – PHASERs Neutron Star Multiverse Slingshot Dynamics Spiritual and UFO Events, and the Multiverse Microscopic Entry into the Multiverse* (Blaha Research, Auburn, NH, 2014).

______, 2015a, *PHYSICS IS LOGIC PAINTED ON THE VOID: Origin of Bare Masses and The Standard Model in Logic, U(4) Origin of the Generations, Normal and Dark Baryonic Forces, Dark Matter, Dark Energy, The Big Bang, Complex General Relativity, A Megaverse of Universe Particles* (Blaha Research, Auburn, NH, 2015).

______, 2015b, *PHYSICS IS LOGIC Part II: The Theory of Everything, The Megaverse Theory of Everything, U(4)⊗U(4) Grand Unified Theory (GUT), Inertial Mass = Gravitational Mass, Unified Extended Standard Model and a New Complex General Relativity with Higgs Particles, Generation Group Higgs Particles* (Blaha Research, Auburn, NH, 2015).

______, 2015c, *The Origin of Higgs ("God") Particles and the Higgs Mechanism: Physics is Logic III, Beyond Higgs – A Revamped Theory With a Local Arrow of Time, The Theory of Everything Enhanced, Why Inertial Frames are Special, Universes of the Mind* (Blaha Research, Auburn, NH, 2015).

______, 2015d, *The Origin of the Eight Coupling Constants of The Theory of Everything: U(8) Grand Unified Theory of Everything (GUTE), S^8 Coupling Constant Symmetry, Space-Time*

Dependent Coupling Constants, Big Bang Vacuum Coupling Constants, Physics is Logic IV (Blaha Research, Auburn, NH, 2015).

______, 2016a, *New Types of Dark Matter, Big Bang Equipartition, and A New U(4) Symmetry in the Theory of Everything: Equipartition Principle for Fermions, Matter is 83.33% Dark, Penetrating the Veil of the Big Bang, Explicit QFT Quark Confinement and Charmonium, Physics is Logic V* (Blaha Research, Auburn, NH, 2016).

______, 2016b, *The Periodic Table of the 192 Quarks and Leptons in The Theory of Everything: The U(4) Layer Group, Physics is Logic VI* (Blaha Research, Auburn, NH, 2016).

______, 2016c, *New Boson Quantum Field Theory, Dark Matter Dynamics, Dark Matter Fermion Layer Mixing, Genesis of Higgs Particles, New Layer Higgs Masses, Higgs Coupling Constants, Non-Abelian Higgs Gauge Fields, Physics is Logic VII* (Blaha Research, Auburn, NH, 2016).

Chrystal, G., 1961, *Textbook of Algebra Part One* (Dover Publications, Inc., New York, 1961).

Eddington, A. S., 1952, *The Mathematical Theory of Relativity* (Cambridge University Press, Cambridge, U.K., 1952).

Fant, Karl M., 2005, *Logically Determined Design: Clockless System Design With NULL Convention Logic* (John Wiley and Sons, Hoboken, NJ, 2005).

Heitler, W., 1954, *The Quantum Theory of Radiation* (Claendon Press, Oxford, UK, 1954).

Huang, Kerson, 1992, *Quarks, Leptons & Gauge Fields 2nd Edition* (World Scientific Publishing Company, Singapore, 1992).

Misner, C. W., Thorne, K. S., and Wheeler, J. A., 1973, *Gravitation* (W. H. Freeman, New York, 1973).

Sagan, H., 1993, *Introduction to the Calculus of Variations* (Dover Publications, Mineola, NY, 1993).

Sakurai, J. J., 1964, *Invariance Principles and Elementary Particles* (Princeton University Press, Princeton, NJ, 1964).

Streater, R. F. and Wightman, A. S., 2000, *PCT, Spin, Statistics, and All That* (Princeton University Press, Princeton, NJ 2000).

Weinberg, S., 1972, *Gravitation and Cosmology* (John Wiley and Sons, New York, 1972).

Weinberg, S., 1995, *The Quantum Theory of Fields Volume I* (Cambridge University Press, New York, 1995).

Weyl, H., 1950, *Space, Time, Matter* (Dover, New York, 1950).

Weyl, H., (Tr. S. Pollard et al), 1987, *The Continuum* (Dover Publications, New York, 1987).

INDEX

About the Author

Stephen Blaha is a well known Physicist and Man of Letters with interests in Science, Society and civilization, the Arts, and Technology. He had an Alfred P. Sloan Foundation scholarship in college. He received his Ph.D. in Physics from Rockefeller University. He has served on the faculties of several major universities. He was also a Member of the Technical Staff at Bell Laboratories, a manager at the Boston Globe Newspaper, a Director at Wang Laboratories, and President of Blaha Software Inc and of Janus Associates Inc. (NH).

Among other achievements he was a co-discoverer of the "r potential" for heavy quark binding developing the first (and still the only demonstrable) non-abelian gauge theory with an "r" potential; first suggested the existence of topological structures in superfluid He-3; first proposed Yang-Mills theories would appear in condensed matter phenomena with non-scalar order parameters; first developed a grammar-based formalism for quantum computers and applied it to elementary particle theories; first developed a new form of quantum field theory without divergences (thus solving a major 60 year old problem that enabled a unified theory of the Standard Model and Quantum Gravity without divergences to be developed); first developed a formulation of complex General Relativity based on analytic continuation from real space-time; first developed a generalized non-homogeneous Robertson-Walker metric that enabled a quantum theory of the Big Bang to be developed without singularities at $t = 0$; first generalized Cauchy's theorem and Gauss' theorem to complex, curved multi-dimensional spaces; received Honorable Mention in the Gravity Research Foundation Essay Competition in 1978; first developed a physically acceptable theory of faster-than-light particles; first derived a composition of extrema method in the Calculus of Variations; first quantitatively suggested that inflationary periods in the history of the universe were not needed; first proved Gödel's Theorem implies Nature must be quantum; provided a new alternative to the Higgs Mechanism, and Higgs particles, to generate masses; first showed how to resolve logical paradoxes including Gödel's Undecidability Theorem by developing Operator Logic and Quantum Operator Logic; first developed a quantitative harmonic oscillator-like model of the life cycle, and interactions, of civilizations; first showed how equations describing superorganisms also apply to civilizations. A recent book shows his theory applies successfully to the past 14 years of history and to *new* archaeological data on Andean and Mayan civilizations as well as Early Anatolian and Egyptian civilizations.

He first developed an axiomatic derivation of the forms of The Standard Model from geometry – space-time properties – The Extended Standard Model. It has a Dark Matter sector that approximates the ElectroWeak sector with Dark doublets and Dark gauge interactions. It also uses quantum coordinates to remove infinities that crop up in most interacting quantum field theories and additionally to remove the infinities that appear in the Big Bang and generate an inflationary growth of the universe. The Extended Standard Model has an ultra-high energy GUT (Grand Unified Theory) limit with a $U(4)\otimes U(4)$ symmetry; and can be united with gravitation to form a Theory of Everything. (See *Physics is Logic Part II.*)

Blaha has had a major impact on a succession of elementary particle theories: his Ph.D. thesis (1970), and papers, showed that quantum field theory calculations to all orders in ladder approximations could not give scaling deep inelastic electron-nucleon scattering. He later showed the eigenvalue equation for the fine structure constant α in Johnson-Baker-Willey QED had a zero at $\alpha = 1$ not 1/137 by solving the Schwinger-Dyson equations to all orders in an approximation that agreed with exact results to 4th order in α thus ending interest in this theory. In 1979 at Prof. Ken Johnson's (MIT) suggestion he calculated the proton-neutron mass difference in the MIT bag model and found the result had the wrong sign reducing interest in the bag model. These results all appear in Physical Review papers. In the 2000's he repeatedly pointed out the shortcomings of SuperString theory and showed that The Standard Model's form could be derived from space-time geometry by an extension of Lorentz transformations to faster than light transformations. This deeper space-time basis greatly increases the possibility that it is part of THE fundamental theory.

In graduate school (1965-71) he wrote substantial papers in elementary particles and group theory: The Inelastic E- P Structure Functions in a Gluon Model. Phys. Lett. B40:501-502,1972; Deep-Inelastic E-P Structure Functions In A Ladder Model With Spin 1/2 Nucleons, Phys.Rev. D3:510-523,1971; Continuum Contributions To The Pion Radius, Phys. Rev. 178:2167-2169,1969; Character Analysis of U(N) and SU(N), J. Math. Phys. 10, 2156 (1969); and The Calculation of the Irreducible Characters of the Symmetric Group in Terms of the Compound Characters, (Published as Blaha's Lemma in D. E. Knuth's book: *The Art of Computer Programming Vols. 1 – 4*).

In the early 1980's Blaha was also a pioneer in the development of UNIX for financial, scientific and Internet applications: benchmarked UNIX versions showing that block size was critical for UNIX performance, developing financial modeling software, starting database benchmarking comparison studies, developing Internet-like UNIX networking (1982) and developing a hybrid shell programming technique (1982) that was a precursor to the PERL programming language. He was also the manager of the AT&T ten-year future products development database. His work helped lead to commercial UNIX on computers such as Sun Micros, IBM AIX minis, and Apple computers.

In the 1980's he pioneered the development of PC Desktop Publishing on laser printers. and was nominated for three "Awards for Technical Excellence" in 1987 by PC Magazine for PC software products that he designed and developed.

Recently he has developed a theory of Megaverses – actual universes of which our universe is one – with quantum particle-like properties based on the Wheeler-DeWitt equation of Quantum Gravity. He has developed a theory of a baryonic force, which had been conjectured many years ago, and estimated the strength of the force based on discrepancies in measurements of the gravitational constant G. This force, operative in 15-dimensinal space, can be used to escape from our universe in "uniships" which are the equivalent of the faster-than-light starships proposed in the author's earlier books. Thus travel to other universes, as well as to other stars is possible.

Blaha also considered the complexified Wheeler-DeWitt equation and showed that its limitation to real-valued coordinates and metrics generated a Cosmological Constant in the Einstein equations.

The author has also recently written a series of books on the serious problems of the United States and their solution as well as a book on the decline of Mankind that will follow from current social and genetic trends in Mankind.

In the past twelve years Dr. Blaha has written over 40 books on a wide range of topics. Some recent major works are: *From Asynchronous Logic to The Standard Model to Superflight to the Stars, All the Universe!, SuperCivilizations: Civilizations as Superorganisms, America's Future: an Islamic Surge, ISIS, al Qaeda, World Epidemics, Ukraine, Russia-China Pact, US Leadership Crisis,The Rises and Falls of Man – Destiny – 3000 AD: New Support for a Superorganism MACRO-THEORY of CIVILIZATIONS From CURRENT WORLD TRENDS and NEW Peruvian, Pre-Mayan, Mayan, Anatolian, and Early Egyptian Data, with a Projection to 3000 AD,* and *Mankind in Decline: Genetic Disasters, Human-Animal Hybrids, Overpopulation, Pollution, Global Warming, Food and Water Shortages, Desertification, Poverty, Rising Violence, Genocide, Epidemics, Wars, Leadership Failure.*

He has taught approximately 4,000 students in undergraduate, graduate, and postgraduate corporate education courses primarily in major universities, and large companies and government agencies.

The above paragraphs summarize much of his work over the past fifty years. This work is fully documented. He continues to engage in research and writing at Blaha Research.

www.ingramcontent.com/pod-product-compliance
Lightning Source LLC
Chambersburg PA
CBHW082008190326
41458CB00010B/3118
* 9 7 8 0 9 9 7 0 7 6 1 4 1 *